Praise for *From Outer Space*

T0274901

"Edgar was a man very willing to go where no one has gone before."

—Avi Loeb, *New York Times* best-selling author of *Extraterrestrial* and chair of Harvard's Department of Astronomy

"Edgar Mitchell has produced a miraculous book. He writes as both a scientist and a spiritual seeker. In sharing from his heart, Dr. Mitchell has provided us with a blueprint for revolutionizing our lives and our planet."

—Wayne W. Dyer, author of *The Power of Intention*

"Edgar Mitchell is one of those rarest of scientists—a true explorer. His journey into outer space has been matched by a lifelong journey into inner space, where he investigated the final frontier, the nature of mind, and returned with nothing less than an extraordinary new science of life."

—Lynne McTaggart, best-selling author of *The Field* and *The Intention Experiment*

"In the history of the human race, twelve people have set foot on the moon. Now one of them has written an important book, an account of a modern-day hero's journey. Astronaut Edgar Mitchell pursued a vision, ventured on a quest, risked great danger, and obtained uncommon wisdom. Now he has returned to share it. We would be wise to listen."

—Larry Dossey, MD, author of *Healing Words*

"Originally scheduled for the ill-fated Apollo 13 mission, Mitchell, as told in this smooth blend of autobiography and exegesis, journeyed to the Moon in 1971. Within a few years, he had left NASA and founded the Institute of Noetic Sciences, aimed at the systematic study of the nature of consciousness. At the institute, he came to some fascinating conclusions, detailed here and based on principles of resonance, regarding a possible natural explanation for psychic powers. Mitchell isn't afraid to go out on a limb; his contention that the universe 'intended' to evolve to higher levels, for example, goes against mainstream Western science. He grounds his ideas in data and reason, however, making this a strong offering for those who enjoy the books of Larry Dossey, Ken Wilber, and others pushing the envelope of the science/spirit paradigm."

—*Publishers Weekly*

"Apollo 14 astronaut Mitchell offers a vision in which technology and intuition are harmonized in pursuit of a more advanced consciousness."

—*Kirkus Reviews*

"Profound and articulate . . . Edgar Mitchell has really dug into the main paradoxical issues [of science and religion] and comes up with a resolving viewpoint. An inspiring piece of literature."

—Harold E. Puthoff, PhD, physicist, Institute
for Advanced Studies at Austin

"Reading astronaut Mitchell's story is like reading a long-awaited letter from a friend—a friend who has journeyed far and wide and seen so much beyond most of us that we read with bated breath. Mitchell's writing is exciting, insightful, and majestic, at once a logical, scientific, and moving spiritual vision. As I read the book, I felt I was looking through his eyes, sensing the unity of the world and seeing our planets and universe for the first time."

—Fred Alan Wolf, author of *The Yoga of Time
Travel* and *Taking the Quantum Leap*

FROM OUTER SPACE TO INNER SPACE

An Apollo Astronaut's Journey Through the Material and Mystical Worlds

DR. EDGAR MITCHELL, SC. D.
with Dwight Williams

Foreword by AVI LOEB
Afterword by DEAN RADIN

NEW PAGE

This edition first published in 2022 by New Page Books, an imprint of
Red Wheel/Weiser, LLC
With offices at:
65 Parker Street, Suite 7
Newburyport, MA 01950
www.redwheelweiser.com

ISBN: 978-1-63748-009-0
Library of Congress Cataloging-in-Publication
Data available upon request.

Cover design by Sky Peck Design
Cover art by Victor Habbick Visions/Science Photo Library/Getty Images
Interior by Steve Amarillo / Urban Design LLC
Typeset in Emtype Akkordeon and Adobe Sabon and Aller

Printed in the United States of America
IBI
10 9 8 7 6 5 4 3 2 1

To my persistent and devoted colleagues of many nations and diverse disciplines who seek to unravel the myths, paradoxes, mysteries, and dogma of the past in order to reveal a path in the modern age toward a sustainable world order.

Contents

Invisible Realities

A Dyadic Model: Interconnections

Portraits of Reality: Interpretation and Paradox

Synthesis

Toward the Future

Foreword

Being the first to do something is often not easy. This is something Edgar Mitchell was very familiar with, as he was a test pilot for the Navy, a founder of Institute of Noetic Sciences, and the sixth man ever to walk on the moon. Truly, Edgar was a man very willing to go where no one has gone before, and he loved to encourage people to follow his steps but also to make their own mark.

Much of his life was spent exploring and supporting others in their work, often delving into what are traditionally taboo scientific areas. Nothing should be taboo from scientific research, and more open doors are needed to encourage scientific exploration to engender breakthroughs of knowledge, which often come from people other than those already well-experienced and known in their field.

The ancient Greek poet Archilochus wrote: "The fox knows many things, but the hedgehog one big thing." When I started my career in astrophysics, the advice I received from my mentor, John Bahcall, was to use the hedgehog strategy in securing tenure: "Focus on one area of research and aim to be the world expert in it."

This was wise advice that accomplished its goal, as I received tenure a few years later. Within two decades, I became the world expert on the first stars in the universe. But despite the success of the hedgehog strategy, I began to develop doubts about it. The doubts intensified when I and my former graduate student Steve Furlanetto wrote an extensive textbook on the emerging frontier that explored the earliest galaxies and the epoch of reionization. Our purpose was to encourage young researchers to enter the field and make new discoveries. Their findings would require us to make substantial updates to future editions of the book, but with the great benefit of learning something new.

Of course, as the field trended in popularity into the mainstream, established cosmologists developed expertise in it too. At that point,

I noticed that these so-called "experts" exhibited the unfortunate tendency to discourage younger scientists who had fresh ideas from entering into their field. The behavior appeared similar to the way animals protect their territory: they wish to remain dominant, minimize competition for available resources, and never expose their weaknesses—in the case of the experts, the important insights they might be missing. The perspective of an outsider poses a threat to conventional thinking. The consequences are particularly acute when the outsider raises foundational questions to which there are no good answers.

The situation is similar outside academia. When a few youngsters from a growing company called Microsoft entered the executive office of Encyclopædia Britannica,Inc., in the 1980s and offered to collaborate on an electronic CD-ROM version of the lucrative encyclopedia, their offer was declined by Britannica's senior management. At that time, the print version of the encyclopedia was a prominent luxury item with an unparalleled reputation. As a result of the rejection, Microsoft released its own digital encyclopedia, *Encarta,* in 1993. Two decades later, Britannica ceased production of its print version, recognizing that it was unable to compete with online resources like Wikipedia. Ironically, the history of Britannica is now summarized on a Wikipedia page.

Is there any evidence that spreading your focus across multiple scientific disciplines necessarily leads to a superficial impact? To the contrary, history demonstrates that polymaths like Leonardo da Vinci, René Descartes, Gottfried Leibniz, Isaac Newton, Charles Darwin, Benjamin Franklin, Marie Curie, and Nikola Tesla were all responsible for foundational breakthroughs in science. Recognizing the added value of the fox strategy, I was motivated to bring together astronomers, physicists, mathematicians, and philosophers under the interdisciplinary umbrella of the Black Hole Initiative at Harvard University. That way, "non-experts" might help "experts" realize what they had been missing on unsolved puzzles about black holes. Knowledge is an island in an ocean of ignorance, and a fresh perspective could identify distant unexplored lands.

When, on a few occasions, I started to suggest out-of-the-box ideas to "experts" in other fields, most of my proposals were flatly rejected. In retrospect, some of them turned out to be right on target. After noticing the experience repeat, I decided to just follow my curiosity and not be swayed away by colleagues who warn, in effect, "Stay out of this lane." I realized that as long as I have tenure and my self-esteem does not depend on prizes or honor societies, I am not sacrificing much by following my own compass. Once an argument is validated beyond a reasonable doubt, the "experts" will agree with it, although they might also claim that they knew it all along.

My advice to young scientists is therefore different from what I received—at least, after you have secured tenure. Define your path not by looking at the surrounding geography and restricting your expertise to intellectual boundaries but by following your internal compass, which is an idea Edgar Mitchell would likely support. Unwarranted resistance by experts might signal that you are on track for an important breakthrough. In the end, wearing early rejection as a badge of honor might give you more pride than any prize awarded to you afterward by the same experts.

Edgar Mitchell was unlike most people. Scientist, Navy officer, pilot, moonwalker, and, most importantly, explorer. We could use more people in the world like him, willing to push boundaries. As Galileo reasoned after looking through his telescope, "in the sciences, the authority of a thousand is not worth as much as the humble reasoning of a single individual." To which I would add the footnote that sometimes Mother Nature is kinder to innovative ideas than people are.

—Avi Loeb,
chair of Harvard's Department of Astronomy,
founding director of Harvard's Black Hole Initiative,
and director of the Institute for Theory and Computation

Acknowledgments

To name individually all who have contributed to this work would require a separate volume. Each person I have met—critic or ardent supporter—has been my teacher. I am grateful for the lessons, even those I did not willingly choose.

I wish to thank those who had faith in an idea that led to the founding of the Institute of Noetic Sciences: Henry Rolfs (deceased) and Zoe Rolfs, Richard Davis (deceased), Judith Skutch Whitson, Paul Temple, Phillip Lukin (deceased), and John White. And those who came a bit later to carry the idea further: Osmond Crosby, Brendan O'Regan (deceased), Diane Brown Temple, Willis Harman (deceased), Winston Franklin (deceased), Barbara McNeill, and the many (and growing) dedicated directors, officers, and staff.

The institute is now recognized, and its agenda studied in more than 30 nations globally. Those who assumed the leadership to carry the ideas to Canada, South America, Europe, and Africa are gratefully acknowledged: in Canada, Diana Cawood; in Brazil, Tamas Makray; in Argentina, Ana de Campos and Ana Lia Alvarez; in England, Ian and Victoria Watson; in Norway, Jonathan Learn; in Nigeria, Hafsat Abiola—to name just a few. Thank you all for your confidence.

Particularly I wish to acknowledge the hard work and dedication of my friend and close associate, Robert Staretz, who has read, reread, edited, made suggestions, and otherwise contributed significantly to this work. Thank you, Bob.

View from the Velvet Blackness

An Inspiration

1

In January of 1971, I boarded a spacecraft and traveled to an airless world of brilliant clarity. The soil there is barren and gray, and the horizon always further than it appears. It is a static world that has only known silence. Upon its landscape human perspective is altered.

During the 15 years prior to the moment my friend Alan Shepard and I opened the door to the lunar module and descended the ladder to the dusty surface of the moon, my days had progressed more or less as I'd planned. But this wasn't the achievement of an individual, a space agency, or even a country. This was, rather, the achievement of our species, our civilization. Life had come a long way since it first sprang from the Earth's rock and water. And now, hundreds of thousands of miles away on that small blue and white sphere, millions of human beings were watching two men walk about the surface of another world for the third time in our history. These were momentous days, extraordinary for their audacity, extraordinary for the coordination of minds and skills that made them possible. A lot of hard work by some of the most brilliant men and women on the planet had culminated in making us a space-faring species. But what I did not know as Alan and I worked on that waterless world, in a mountainous region known as Fra Mauro, was that I had yet to grasp what would prove most extraordinary about the journey.

It wasn't until after we had made rendezvous with our friend Stu Roosa in the *Kittyhawk* command module and were hurtling Earthward at several miles per second, that I had time to relax in weightlessness and contemplate that blue jewel-like home planet suspended in the velvety blackness from which we had come. What I saw out the window was all I had ever known, all I had ever loved and hated, all that I had longed for, all that I once thought had ever been and ever would be. It was all there suspended in the cosmos on that fragile little sphere. What I experienced was a grand epiphany accompanied by exhilaration, an event I would later refer to in terms that could not be more foreign to my upbringing in west Texas, and later, New Mexico. From that moment on my life would take a radically different course.

What I experienced during that three-day trip home was nothing short of an overwhelming sense of universal *connectedness*. I actually felt what has been described as an ecstasy of unity. It occurred to me that the molecules of my body and the molecules of the spacecraft itself were manufactured long ago in the furnace of one of the ancient stars that burned in the heavens about me. And there was the sense that our presence as space travelers, and the existence of the universe itself, was not accidental, but that there was an intelligent process at work. I perceived the universe as in some way conscious. The thought was so large it seemed at the time inexpressible, and to a large degree it still is. Perhaps all I have gained is a greater sense of understanding, and perhaps a more articulate means of expressing it. But even in the midst of epiphany I did not attach mystical or otherworldly origin to the phenomenon. Rather, I thought it curious and exciting that the brain could spontaneously reorganize information to produce such a fantastically strange experience.

By the time the red-and-white parachutes blossomed in the life-giving atmosphere of Earth three days later and our capsule splashed into the ocean, my life's direction was about to change. I didn't know it then, but it was. What lay in store was an entirely different kind of journey, one that would occupy more than 40 years of my life. I have often likened that experience to a game of pick-up sticks: Within a few days my beliefs about life were thrown into the air and scattered about. It took me 20 years to pick up those sticks and make some kind of sense of it all, and I now believe I can describe it with an adequate degree of comprehensibility and scientific validity. I like to think that this book is the result of both journeys.

Shortly after returning from the moon I was often invited to speak at various occasions. In lecture halls and auditoriums across the country two questions were inevitably asked. The first was, how do you go to the bathroom in space? The second was, what did it feel like to walk on the moon? The first was usually asked by children because they really wanted to know, and are less inhibited than adults. The second quickly became irritating simply because I didn't know the answer. It was certainly a sensible question—I was an astronaut, after all, one of 12 men to have walked upon the surface of the moon. People would naturally want to know. But when I finally asked myself why the question was so bothersome, it occurred to me that there were emotional realms lodged deep in my own psyche that I hadn't fully explored. I now find it interesting and a bit amusing that it bothered me at all. But it did, and for a very particular reason: Somehow I couldn't resurrect the feelings I had while there, though my thoughts and actions were easily summoned.

Years ago I began my flying career as a Navy pilot. On heaving black seas in the middle of the night I have landed large jet aircraft on rather small, converted World War II aircraft carriers—a situation in which, quite literally, your life depends on the cumulative experience you have acquired through many years of practice. It was intuition you depended on, it was *feel* adding to normal sensory data with which you guided your aircraft as you carefully tried to avoid a collision with the deck. But it wasn't a feeling that created emotion in the moment. Out of necessity, emotion had to be suppressed. What I lacked in my early years was an understanding of how intuition, emotion, and intellect all interrelate.

Not long after entering the lecture circuit, I asked two friends, Dr. Jean Houston and Robert Masters, to regress me under hypnosis so that I might learn a few things about myself. I wanted to know both why I didn't remember my feelings while on the moon, and why the question irritated me so much. Ultimately I wanted to understand what psychic-sensitive and highly intuitive people were aware of, and what they experienced. But first I had to examine myself—to examine all my wants and needs and flaws, and honestly describe myself to the point that I could say, yes, I am like that. Thus began an arduous study of my own inner experiences.

After leaving NASA in 1972, I founded the Institute of Noetic Sciences in California. This would fund much of the scientific research that I wanted to see accomplished to help me better resolve the complex insights from my experiences in space. Since then the institute has thrived, but it has been a bit of a challenge, at times, to keep it from becoming a church. Some of the folks I've come across in my lifetime have held some eccentric and dogmatic beliefs about space, the cosmos, and the ultimate nature of reality. And on many occasions it has seemed as though I was expected to become a high priest in some kind of new religion. Frivolous connections were made between the fact that 12 men walked on the moon and that there were 12 disciples of Jesus. Furthermore, I wore a beard at the time, and the absurdity seemed to expand into the messianic realm. So I shaved the beard. From the very beginning I realized I had to be suspicious of everything I heard, and everything I knew—or even thought I knew. It was of particular importance to retain my individuality and not become enamored with any particular established thought structure along the way. Evidence would set the direction. I came to recognize the effects of my own belief system and the powerful role of enculturated belief systems in general; I needed to reexamine accepted thought with new eyes.

To those around me at the time, I suppose I would have seemed a rather unlikely candidate for this second journey. During the Korean War I was a Navy pilot, and for some time afterward a test pilot. After the flight of Sputnik in 1957, I chose to alter that course and sought a role in the space program. The training required for a jet pilot and astronaut is somewhat incompatible with that required for a modern-day Shaman. And that's more the way I saw myself as I settled into this new journey, and how I see myself today.

This is not merely a romantic idea, but rather the role I have chosen as an explorer to better understand the universe, having had the privilege of seeing it from an extraterrestrial point of view. Though the course of the journey has turned me inward, I've tried to retain my scientific sensibilities. My life's purpose, I now see, has been to discover, to reveal, and interpret information, first in outer space, and now in inner space. I have always dealt in the here and now in a meat-and-potatoes sort of way; I've wanted to solve problems simply because they were there and were intriguing.

This is all by way of saying that the purpose of this latter journey has not been to form another cult (the world has plenty), but to reveal more accurately and more fully the structure of reality as we experienced it in the late 20th century as an emerging spacefaring civilization. When I returned from the moon I saw perhaps a little more clearly how our traditional modes of understanding did not adequately explain our modern-day experience. We needed something new in our lives, revised notions concerning reality and truth. Most of us have accumulated this body of ideas that make up our belief system through external authorities rather than through our own quest and original insight. Our beliefs were, and still are, in crisis.

To have lived in the 20th century is to have witnessed the extraordinary miracle and folly of humankind firsthand. There hasn't been a century that approaches the height of its achievements nor the depths of its mayhem and despair. Ours has been a century of demystification, manmade miracles, and man-made catastrophe—most never previously thought to be possible. And those of my generation have perhaps seen the most. We have seen the world evolve from the simple, gray years of the Great Depression and World War II, through the incandescent Nuclear Age, born over the glassy sands of the American West where Poncho Villa and Butch Cassidy roamed on horseback just a few decades earlier. Progress has been swift and severe. We've lived through the silent terror of a war that was never fought, then presided over one fought over ideology against a proxy opponent in a distant jungle, and are now engaged an era of organized global terrorism the likes of which has never before been experienced. We have seen men catapulted into outer space without knowing what they would find there; we've seen men climb mountains of the moon, where they beamed the picture back to millions of magic boxes in living rooms, taverns, shops, and kitchens around the world. Whoever said the Age of Miracles passed long ago hasn't been paying attention.

What our children won't see is the trajectory of this evolution, its defining arc; that story must be recorded for them in the living pages of books or other media. There will be no horse-drawn plows or penny farthings in the coming centuries. Our lives will eventually pass, recorded only on celluloid and the page, silicon, or digitally, as a kind of artifact—cave drawings from the 20th century. Religions of the world will loom, then fade—or remain, depending upon their ability to adapt to the everchanging notions about reality they were created to discover.

Even in our time, we still cling to the idea of the supernatural, the demonic, the divine. We use it when science seems to offer no acceptable explanation. In medieval times there was no science, only religion. Since Rene Descartes, each belief system has been allowed to proceed down a separate, noninterfering path. And for 400 years they have enjoyed a peculiar independence, as Descartes believed thought and matter were of two different realms. This dualistic philosophy has allowed Western science and religion to evolve as we now know them. The Church has left science to the scientist, the scientist has left religion to the theologian, and they have more or less peacefully coexisted (with a few notable exceptions) ever since.

It isn't an overstatement to say that Descartes opened the way for Newton and the early classical scientists, then much later Planck, Einstein, Bohr, and finally the new model of quantum mechanics. It is this revolutionary scientific model that finally penetrated the veil of religion by demonstrating that the act of observation could affect the observed. These realms of thought must not merely coexist in the mind of the scientist and the theologian, but must be allowed to become integrated—simply because they are so obviously intertwined. Sooner or later this reconciliation is inevitable, as the scientific method has shown itself powerful enough to discover its own flaws. I have come to believe evolution has progressed such that we must now assume a large measure of conscious control in our own evolutionary process, as human volition is in fact a fundamental characteristic of nature.

I am one of a growing handful of human beings to have seen the Earth from the point of view of an extraterrestrial. In the heavens there is no up and down, no east or west. Earth is but a beautiful blue speck in the midst of a vast emptiness marked by luminous celestial bodies. We inhabit but one of those celestial bodies; one of the most organized—for all we know. From the heavens, in 1971, the Earth looked peaceful and harmonious, but of course all was not as it appeared. Conflict that threatened our very survival lay below. Weapons were poised, ready to annihilate life as we knew it at a moment's notice; environmental crises were lurking just beyond public awareness. The common root of these mushrooming dilemmas, I believe, has been conflicting, outdated, flawed ideology and dogma, with roots in antiquity.

It has occurred to me that human destiny is still very uncertain, that the veneer of civilization is yet exceedingly thin, and our current actions are not sustainable. Believing as I do that the universe is an intelligent system, and understanding the absurd and tragic fate that may await us, I have wondered if we are prepared for our own survival, if our own collective consciousness is yet highly enough evolved. Our universe seems to learn by the blunt process of trial and error. But I now understand that we have a certain degree of control over the evolutionary process and can influence our own course. But the only way to accomplish this is by bringing into question the very way we think about consciousness and the universe; by questioning many fundamental assumptions underlying civilization.

This is a challenging story, one that requires a certain dedication on the reader's part, as it contains thought from various scientific and religious disciplines. That, in fact, is at the very core of this book: a synthesis of scientific and religious modes of thinking, a movement toward the creation of commerce between the two so that the structure of the universe itself is more fully revealed. But I think it is first necessary that I tell you something of myself, and in so doing, reveal my motives for the unusual course of my life—I should say, my two lives. The first I now see was spent in the interest of taking a physical journey, while the second has been consumed by a spiritual and intellectual quest. It has taken both, I believe, to arrive at the conclusions I've drawn from the sum of my experiences concerning the nature of reality. The results I have fashioned into a model, a *dyadic* model that describes the universe I experienced as accurately as anything I can come up with.

The narrative is not meant to be pedagogic, and my conclusions are only based upon a proposed model of reality that I believe deserves wider consideration, and which, since this work's initial publication, have received substantial validation. The book requires a degree of open-mindedness and a willingness on the reader's part to investigate abstract realms of thought and arcane ideas. Perhaps above all else, it asks the reader to see himself or herself as a part of an evolving universe, and as an extraterrestrial, just as I saw myself when I gazed about, suspended in the heavens almost 40 years ago.

Sea of Grass

The Early Years

2

My mother wanted me to be either a preacher or a musician. She was an artist by temperament and a farm wife by necessity. She didn't see much benefit in the making of war, and I suppose I've never forgotten that. I also suppose I've tried to resolve this conflict from the very beginning.

I was born into what had been a prosperous ranch family in the midst of the Great Depression and in the Dust Bowl. Lives tended to be brutish and short here on the plains of West Texas where the individual seemed so exposed to the harsher acts of nature. Daily life was primitive, and these were especially difficult times. The Southern Baptist faith of my mother and grandmother provided the hope that with hard work, prosperity would return. As my father would say, we were not poor—only short on cash.

My random childhood memories are happy ones. They are scenes of wide-open spaces with shanties leaning against a constant wind, wheat fields heaving beneath a vast sea of sky. Three years after the wheat crops failed, the men-folk drove spikes on the Santa Fe Railroad, and our family retreated to a piece of land on which stood a three-room clapboard structure with outdoor plumbing. The men and women alike saw this as temporary privation to be endured.

A small creek cut through the prairie, providing good bottom land on a portion of the farm. That year my father grew a crop of cotton on its 40 acres, while my mother raised my infant sister Sandra and myself. In the front yard was parked a worn 1929 Buick coup with a rumble-seat, and in the back stood the tack shed and corral with plow horses and a milk cow. The following year my grandfather reassembled the family in Roswell, New Mexico, near where the Lincoln County Wars were fought a few decades earlier. Through astute trading he gradually acquired a small but growing herd of cattle that put the family back in business.

From the center of town the echo of hymns could be heard on Sunday morning as they issued out from under a white steeple. Perhaps I was taught the fear of God in this setting because it seemed obvious that there was so much to fear. But I grew out of a tradition of self-reliance and trust in one's instincts, those mythic values of the Old West, and as I came into adolescence I suppose it was natural to question precisely why we should live so fearfully. My grandfather certainly did not. He was known around these parts as Bull Mitchell because of his livestock-trading acumen. Some of the ranchers who would later grow wealthy (though you wouldn't know it by their bent and sweat-stained Stetsons) often went down to Argentina or Brazil on business and sent back postcards with no address other than "Bull Mitchell, New Mexico." And he would generally get his mail. That is to say, he was widely known, but to a child he was immortal, bullet-proof. Above all else, he was fearless.

From the very beginning I naturally gravitated toward the male side of the family. I have memories from when I was a young boy of a trail of red cedar shavings strewn along the concrete sidewalks of Roswell that I would follow this way and that, drawn along by the magnetic pull of my grandfather as he whittled through a lazy afternoon. He would casually move here and there, whittling, wherever conversation with friends or a cattle deal led him. He was the center, and seemingly the originator, of his own universe.

I recall the wide-bodied car he owned in later years, a 1946 Ford, and how he drove it between stationary objects. The once-proud, bulbous fenders were wrinkled, crimped from my grandfather's habit of driving through narrow spaces where only a horse could pass. But he felt entitled to go wherever he chose, and that the car was obligated to take him. The condition of its body spoke volumes of his nature: a 19th-century man set in the vertiginous 20th century; a man born out of time.

Just a mile or so down the road from where I was raised lived a man I imagine was not unlike my grandfather—a man who is now considered the father of rocketry. This was deep in the bleakness of World War II. Across an ocean, his German successor, Werhner von Braun, was busy designing the rockets known as the V-I and V-2, which were arcing across the English Channel and detonating when they collided with downtown London. Each day as I walked to school along the white gravel road I would pass the quiet country home where a mad scientist was said to live. He was, quite literally, a rocket scientist. He was also America's first, and his name was Robert Goddard.

The house was generally quiet. He had recently moved from Massachusetts (some say he was invited to leave), and now worked and studied in austere isolation—far from sensitive populations and their demands for quiet and safety. Though I have no recollections of rockets flaring into the night skies or the ignition of exotic new fuels, there were stories that circulated among the natives of Roswell—stories of fire and brimstone igniting the heavens, strange machinery, and a quiet, reclusive mind assembling it all. This was a man who would loom large in my imagination, a man of the proportions of my grandfather. He was mythic, and I now see how his life ran so counter to the setting he must have found himself in. Here was a man of science, a man from that vast ungodly world beyond the perimeter of Roswell. By any standard, Robert Goddard was part of the scientific lore of the times. Whenever I walked by his farm it was always quiet, yet Wernher von Braun's rockets would continue to terrorize Londoners. The efforts of both would lay the groundwork for what was to occur after the detonation of a terrible new weapon, just outside another small town in New Mexico, that produced strange cerebral clouds in the vast distance. I recall what was the luminous glow from early tests at the White Sands Proving Grounds of the bomb that would bring the war to an abrupt end, and initiate another quiet, cold one. Of course, all of this was well beyond my field of understanding and experience at the time, and unbeknownst to me then, I would one day be very grateful for the work of my neighbor.

When I was 13 I took an ad hoc job at the local airport washing fragile airplanes made of light framing and lacquered cloth. When I was 14 I soloed in one of those planes myself, and experienced for the first time the sense of freedom found only in the seat of an airplane: release from the Earth. And it was during this time in my life that I developed my own interest in science. Early on I sensed I was an engineer by nature. I came to understand farm machinery, as explained to me by my father, and airplanes, as explained by local pilots and mechanics. I came from a self-schooled, intelligent lineage that wanted me to have the finest education affordable. Because we couldn't afford much, sacrifices would be made. I was also blessed with the attentions of rural school teachers who devoted special care to a student they believed would one day leave this town. And in 1948, that's what I did.

The administration at Carnegie Mellon must have thought a cowboy from New Mexico would make an exotic addition to their student body; I probably was, in that Pittsburgh society. Though I didn't excel my first year, I was a serious student. At times I would see myself as I believed others might see me: a cowboy with jug-handled ears and straw in his teeth; simple, but earnest. And every now and then I played to their expectations.

I used time efficiently in those days. I carried an extra course load, pledged Kappa Sigma, met and dated my future wife, Louise, and when I began to run short of funds, took a full-time job in a steel mill, cleaning slag from burned-out blast furnaces. With an equally impecunious friend I pulled wooden clogs over my shoes, then climbed inside the cooling cavern for but a few seconds to huck out hot chunks of slag from its black stomach. After our midnight shift we headed back to the fraternity house to get some rest, and then moved on to class. So that I wouldn't have to do this for very long, I accelerated my undergraduate studies and finished in three and a half years. As soon as I did so, I married Louise and moved back to the ranch in New Mexico, which had by then grown by two farms and two farm machinery dealerships.

These were challenging times for a young man. The conflict on the Korean Peninsula was heating up, and it was made clear that you could either enlist or be drafted. Although military life was not in my career plans, it was unavoidable at the time. I wanted to fly, and as a married man the only way to do so was with the Navy, so I enlisted. Consequently, Louise joined me in San Diego during my final days of boot camp, and then we found ourselves again heading east; our destination was the Officer Candidate School in Newport, Rhode Island. We stopped and visited both sets of parents along the way, and when we arrived in Newport on Christmas Eve, 1952, with wonderful gifts from our families still unopened, we had but 25 cents in nickels and dimes in our pockets. With it, we bought and shared a hot dog and a cup of coffee, and then drove directly to the OCS headquarters where I could collect my first paycheck. That's how it was for us in the beginning: austere and simple, but infinitely hopeful. In spite of present hardship, the future spread itself out before us in a succession of pleasant vistas.

Not long after our arrival, Louise took a job as an instructor seamstress for Singer, and soon discovered she was pregnant. She did her best not to show, as in those days it wasn't uncommon for a woman to lose her job when expecting. But together we survived the 16 weeks of my training and her work, and were off yet again to another part of the country where neither of us had ever been—this time Pensacola, Florida. But now I was an officer and a gentleman with a bit more green in my pocket to support our family.

We drove through the May heat of the South with Louise pregnant and all our belongings piled in the back seat. The day of our arrival we learned to show up at the pressroom of the local newspaper at dawn, where we could rifle through the classifieds for an apartment to rent. This was a training base in wartime, with hundreds of young couples not unlike ourselves searching for some sort of home. But within a few days we did find a modest place. A few months later, in the deepest heat of that unrelenting southern August, a daughter was born to us. We named her Karlyn Louise. Suddenly we were no longer just a couple, but a family. This was about the time that Louise found her life gradually growing emotionally disheveled. I had just begun my pilot school training, a process that would begin with aircraft driven by propeller, then the jet, and finally the rocket. And through it all Louise generally quieted her concerns; either that, or I was too wrapped up in what was happening outside the realm of home and family to notice.

From the very beginning I was drawn to the cutting edge of flight technology as though by some mysterious force. And I was welcomed there. I suppose this was, in part, because I was naturally good at it; I could feel my way in the seat of an aircraft. There was a special sense in flying, as though the aircraft were an extension of my body, which made me stand out as a pilot. It lent me the perception that there was some larger purpose that I was fulfilling, which was of course immensely satisfying. But for a pilot's wife, the lifestyle can be uneasy if not terrifying.

I began my Navy career in the seat of an AT-6, which was then the Navy's standard trainer. Not for some time would I climb into the cockpit of my first jet aircraft. Looking back on it now, the country seemed young and new then, with the advent of nuclear technology, the jet engine, and rocketry. The world itself seemed bright and colorful, poised for the extraordinary. Spaceflight was still only the dream of a handful of scientists. But when I did fly that first Navy jet, I knew this was where I was supposed to be.

After my training in Pensacola, Louise, Karlyn, and I were again on the road, first to an advanced Navy training program, and then to the West Coast, where I'd be flying a large propeller aircraft, the P2V, for the next 18 months. We purchased our first home on the rocky fog-shrouded slopes of Whidby Island, Washington, and settled in for our three-year stay. At the time, military life suited me insofar that it allowed me to fly. And I knew that aircraft and spacecraft would be the next medium of man's exploration, just as the horse and ship had been in the past. Even as a young man in my early 20s I suppose I secretly held the dream of being one of those explorers.

For three years in the Pacific I flew various aircraft that would play small roles in the Korean conflict and the Cold War. It was a tense time, not only for Americans in general, but for those young men and women sent off to that rather obscure corner of the globe. Toward the very end of my first overseas assignment, our plane was attacked by interceptors while on a routine night patrol near the city of Shanghai. The radar operator on my aircraft informed us that two jets were fast approaching in attack position. Somehow I anticipated what was about to happen— perhaps by some sense of intuition—because just as I thrust the controls forward, sending the P2V into a dive, my copilot and I saw the incandescent tracer bullets from one of the jets arc overhead. To call it a close call would be to minimize the threat. But that's how it was for many of the men around me in their daily routine. Life in the military can mean relying on that vague faculty of intuition on a regular, if not daily, basis. This seemed to represent the fundamental nature of the times; it was as though all events occurred in rapid succession, and it took instinct to navigate the course of a life. Immediately upon my return to the United States, I was transferred to carrier duty in San Diego, and a new assignment. Louise had hardly finished making our house into a home when we were once again on the move.

3

Much about military life was troubling in those days. During my three-year stint of flying from various land and carrier bases in the Pacific, I saw what was widely believed to be the future of warfare. From test sites on a small group of islands called Kwajalein Atoll in the South Pacific, I studied the detonation of atomic weaponry, the successor technology to that which produced the extraordinary glow in the skies over White Sands Proving Grounds during my boyhood. The sight of the opaque mushroom cloud and the palpable release of so much energy was at once terrible and awesome. Considering the political landscape of the times, the escalation of a cold war over ideology made for apocalyptic scenarios. This was the ugly aspect of my profession. From the very beginning I hoped it would someday become obsolete, but there was a job to do, and I would do it with all the vigor I possessed.

In 1957, after spending a year flying from carriers in the Pacific, I was sent to an isolated military base just north of the Mojave Desert. The mission was to design a new delivery system for atomic weaponry before the days of cruise missiles and ICBMs (intercontinental ballistic missiles). Above this lonely desert outpost I carried what are known in military circles as "shapes" in the bomb bay of my jet. These were facsimiles of the shape, size, and weight of actual nuclear weapons. As my jet raged through the pale desert sky at maximum speed, 50 feet above the barren dusty floor, I helped perfect the means by which such a bomb could be smuggled below enemy radar, then lofted into the air, detonating after the pilot had raced away, safely ahead of the devastating shock wave that would follow with angry revenge.

Through the course of the hot, unchanging seasons, we carried out our duties in short-sleeve tropical uniform. We were test pilots, absorbed in our work and oblivious to the barren landscape around us. My young family lived on base with no real civilization for hundreds of miles in any direction. There was only the base, surrounded by thousands of square miles of desert. On late afternoons, husbands and wives would often-times congregate on someone's patio around a barbecue pit, talking shop, about the sameness of the weather, and what they might do with a weekend up to Reno. The men and women I was surrounded by were passionate about their work, and because of this, they were simply excit-ing to be around. In this respect I fit in, as this was serious research and development; we were in the process of making aircraft do things they weren't designed to do. But occasionally a pillar of black smoke, rising from the desert floor, would signal that someone might not be coming home that night.

The larger mission of military life never appealed to me. I could manage interest in the details of the weapons technology so that I could carry out my duty. But the horrendous destructive power of the weap-onry was distressing, and I wanted to shift my sights from the technol-ogy of war to that of exploration. I felt that exploration was somehow at the core of my being. I knew that after this duty I might be able to lead my career in another direction by applying for the astronaut corps, though I hadn't yet accumulated enough hours in the seat of a jet. When the Soviets had launched Sputnik in 1957, I realized that humans would not be far behind. I wanted to go there.

But before leaving China Lake, another child was born to us, a baby girl we named Elizabeth. The family was growing, and we were again on the move, this time to the U.S. Naval Post-Graduate School in Monterey, California. Here I would study aeronautical engineering for the first time in my life, incorporating what I experienced in the cockpit into the hoard of new technical concepts being tested in wind tunnels. I also elected to study subjects outside the usual domain of a test pilot's curriculum, such as the Russian language—a sign of the times. This was the confluence of science and real life.

Through these years it became clear to me that new technology would play the preeminent role in the exploration of space, and that being a pilot alone, even a test pilot, would not take me there. I had to become a rocket scientist, like Robert Goddard. When school in Monterey drew to a close, I looked to the East Coast for an additional curriculum in aeronautics and astronautics, and found one to my liking at the Massachusetts Institute of Technology. The program's ultimate goal was in getting a man into outer space. This would lead to the series of missions that would eventually land a man on the moon and return him safely to Earth. By changing my status with the Navy to that of an aeronautical engineer, I found a career path that I hoped would work for myself, my family, and the Navy. So, in 1961, Louise and I gathered our young family, and we were once again embarking on the familiar routine of moving and making a home in a distant city, this time in West Newton, Massachusetts, a middle-class residential community just outside of Boston.

The new program focused on a varied range of subjects never before studied together. These were bold new topics such as star evolution, galactic evolution, optimization theory, orbital mechanics, space navigation, rocket propulsion, inertial platforms—abstruse subjects at the time, often utilizing primitive, simplistic new theories. No one knew the answers to the questions they posed, just as no one knew what we would encounter in the heavens. There were other newly formed curricula similar to this one at Caltech and Princeton, but by and large, this was a new science, and in retrospect, our understanding at the time was rather naïve and touching. Was there life out there? And if so, was it benevolent, or malevolent? Was it primitive, or intelligent and evolved to any meaningful degree? Only the most daring or reckless asked such questions, as conventional wisdom, derived both from science and theology, firmly believed the Earth to be the biological center of the universe. This naïveté was reflected in the popular culture of the times as well. The movies of the 1950s and early '60s are interesting in the way they depict America's childlike understanding with unobscured honesty, along with a touch of fear, just as it really was. In the common mind, heaven was just out there above us, not to be discovered in this life. It's easy and self-indulgent to laugh now, but there was no advantage of hindsight, no view to be had but that brilliant light of discovery, the source of which lay years in the future. I recall that when I was a teenager, an alien spacecraft was announced to have crashed near our ranch in Roswell. A flurry of speculation and gossip followed in its wake, until the next day, when the official explanation was amended and the government cited the incident as having been a fallen weather balloon. Was it? More recent events and disclosures reveal the weather balloon explanation to have been silly and deliberately misleading. These events were in part what made the times so unbearably exciting.

It was here at MIT that I began to learn about quantum mechanics and the theory of relativity in an academic forum. The universe was starting to be explained in ways quite different from the conventional wisdom of classical physics, and certainly different from the way we perceived it to be through our senses. I understood that if a person were to one day go to the moon, the voyager would come back a few microseconds younger than a twin who had stayed behind. I didn't yet fully understand that at the bottom of all matter there was no simple, coherent picture of physical reality. Though they taught the mathematics of the new physics, the larger issues and the deeper implications of the theories had not, even then, trickled down to the ordinary engineering professor. Coming to fully understand the implications of these notions was years away. Only now, as we begin the 21st century is there a wider knowledge and new evidence that not only reinterprets much of 20th-century science, but also shows some of its basic flaws.

In the towns we visited through these nomadic years, Louise and I sought out local parishes where we found pastors and congregations with sensibilities more or less compatible with our own—that is to say, concerned with the deeper questions of purpose, destiny, and the human condition. In West Newton, we found this companionship in the company of Jim Remington, a young married man my age who pastored the local Baptist church. On warm summer evenings Jim and I would sit out on the patio and become lost in discussions that meandered from free will and determinism to the nature of God, to the structure of the universe itself, and how this new science I was becoming acquainted with illuminated our understanding. Jim was an uncommon man of his vocation, in that he was vitally interested in discussion that challenged Christian dogma. He was never dogmatic in his own beliefs, having an intellectual bent that I liked to think was similar to my own. I was coming to realize as an adult that it was possible to place myself within the same sphere of influence as that of a Baptist minister and still maintain an independence of mind. And perhaps as a result I came to realize how my early fundamentalist upbringing had shaped my thought, though I was fast on my way to becoming a dyed-in-the-wool agnostic. My mind, I knew, was somehow capable of containing this paradox. Yet for many years I would harbor some hidden fear of God, some fear brought to bear by the residue of that upbringing, a soft whisper warning me of the eternal penalty for blasphemy or casual irreverence. There was the sense that the potential for damnation lay in the spoken word, and even perhaps in thought.

But this was my own shadowed background, one I ignored as Jim and I sat in the cool evening air making rather erudite investigations into what our place was in this wide world. The climate of inquiry, I silently told myself, was sufficiently reverent for any kind-spirited Creator.

I haven't always had an interest in the details of enigmatic scientific problems. When I came to MIT I was a passable mathematician. The abstract topography of science was a landscape I was comfortable moving about, yet there were deep domains with which I struggled. But here I was at one of the most prestigious institutions of higher learning on the planet, working on a project mandated by a president slain while I was in residence there, all of which was in the interest of landing a man on the moon. In spite of this, I had no interest in exploring the finer, capillary regions of science, but preferred rather the broader issues of morphology and cosmology. My interests lay in how things fit together. I saw mathematics simply as a tool, a language, but one of various methods for obtaining a description of reality in order to better understand it. Reality was there, somewhere, in the form of an intricate landscape to be explored and understood. But I came across those who thought the exactness of the mathematical approach was actually *embedded* in nature. They believed that only by reducing nature to its bare elements and uncovering the inherent mathematics could one gain true insight into the territory they chose to explore. Only the mathematics could point to the next step. These were brilliant scholars, but we stood at opposite intellectual poles. There seemed to be something stirring my intellectual vision, perhaps my own intuition guiding me in another direction. What came to me most naturally was an understanding of synthesis, the relationships and patterns existing across disparate disciplines of thought, not the tedious process of splitting intellectual hairs.

Life in these days was stimulating and varied. At night I would come home from campus, weary but happy, to a young and noisy household.

In the last month of summer we retreated to the coast of Cape Cod, and in the winter we all would ice skate upon a makeshift pond I fashioned out of lengths of wood and plastic, and filled with water from a garden hose (the local ponds for skating were considered too dangerous for children). Once a month I headed out to a Navy air field just south of Boston and climbed into the seat of a plane in order to maintain a certain level of proficiency. By dusk I was back in the lap of domesticity. These were good times in almost every way, but Louise and I were having more difficulties, as the silent traumas of being the wife of a test pilot who aspired to be an astronaut were starting to mount. And I was too ignorant, preoccupied, and simply too insensitive to understand her plight. It never occurred to me, as I forged my own role in this male-dominated world, that I was displaying many of the rigid characteristics of my patriarchal grandfather. My military service had been the result of circumstance, but I'd unexpectedly found a home there. Louise faithfully and lovingly supported my ambitions and stabilized the family. Such dreams might lead to glory or widowhood, and she and the children were silently in need of stability and tranquility.

Precisely what an astronaut did and what he would face in the vast frontier of space was still a looming unknown, not only to the public at large, but to those of us in the space industry as well. The challenge was exhilarating, but the sheer lack of convention in our lives was at times a burden. We had never been firmly grounded in any of the communities we'd breezed into and out of after our brief tenure had expired. We were always off to another distant setting so that I could further my understanding of what I might find, were I ever to make that climb into space one day. I was pushing my family from one venue to the next so that I might fulfill the most exotic of dreams—that of being an explorer in the latter half of the 20th century. What seemed so strange and intoxicating was that it was becoming increasingly clear that this dream was drifting within my grasp. All that was required was a lot of control over the design of my study, an extraordinary degree of flexibility, hard work, and a reasonable dose of luck. But there were never any guarantees, and for a mother and her children, the lifestyle meant friends and playmates left behind.

As my studies in Cambridge drew to a close it was once again time to contemplate what the next leg of this journey should be. There were lots of roads, but no sure map. Eventually I applied for a position in the guidance and control division of NASA, simply because I knew this could take me nearer to space, and a few weeks later I was accepted. In the spring we were again packing our lives into another vehicle and on our way across country to Houston, where the agency had been firmly ensconced by the political acumen of President Lyndon Johnson. Houston was the magical city if you aspired to be an astronaut. But as the family rolled southwestward, the larger direction of the journey was abruptly interrupted when I received a telegram while laying over for a visit at Louise's mother's home in Pittsburgh.

Over the phone I was informed that our destination had been changed from Houston to Los Angeles, where the military's own space program was based. An officer by the name of Captain Jack Van Ness, the head of the Navy liaison office, wanted a deputy with test pilot experience and a strong technical education. Our mission would be to help develop a manned orbiting laboratory that could be used for military surveillance. The Pentagon foresaw valid military applications in space, though Eisenhower's intention two administrations before had been to send men there for the sole purpose of peaceful exploration. During the 1950s the military had campaigned long and hard to control the development in this new frontier, but President Eisenhower wisely placed the new National Aeronautics and Space Administration under the auspices of civilian leadership. The military chiefs, however, continued to find allocations from Congress for their own pet projects in space, and as far as the chain of command was concerned, I was a part of their program, not NASA's.

When we bid farewell to Louise's widowed mother the following day, we were on our way to Los Angeles instead of Houston. For five days we drove over the undulating prairie, into the west, where we set up another life not unlike the one we had just left behind in the east. For a year and a half I worked as Jack Van Ness's deputy, managing and coordinating the design efforts for the orbiter craft and its sensors. It was an exhilarating task, trying to bring together the work and talents of so many different people in the common effort of building something as innovative and lofty as a manned laboratory that would one day orbit the Earth. Its sensors were required to push the state-of-the art in surveillance. But a manned orbiting laboratory was still just a wild idea in the public consciousness.

After a year and a half of hard work, I realized that mere management skills would not take me any closer to that chalky white world in the heavens. Astronaut candidates for the Manned Orbiting Laboratory (MOL) had been picked mostly from the ranks of Air Force test pilots. The only two Navy men selected were hot fighter pilots considerably younger than myself.[1] I knew that if my training was lacking in any one element to a degree that could keep me from going to the moon, it was the lack of time I had spent in recent years flying operational jet aircraft. I prevailed upon Jack Van Ness to help me get assigned to the Aerospace Research Pilots School at Edwards Air Force Base in the Mojave Desert, which was then headed up by none other than the legendary Chuck Yeager.

Here there was an endless list of exotic flight programs. This was the place where the most audacious aircraft in the world were flown to their very limits every day, and to be invited into the astronaut corps, it didn't hurt to be a part of this exclusive fraternity of pilots. Even better, my family could remain in Los Angeles, while I commuted to and from the base on weekends. My children could stay in the same schools for a change, and Louise could make herself more of a home.

The test pilot school at Edwards had recently added a space training curriculum. These were transitional days for those in the testing fraternity, as there had been words between the older stick-and-rudder pilots and those of the new technological hybrid that would be needed not only for computer-aided flight in aircraft, but also for missions into deep space. Old-school pilots, great as they were, clung to the integrity of aircraft with mechanical aerodynamic controls that performed by reacting to the forces of nature within the atmosphere. Now here came these new pilots who aspired to one day "fly" beyond the atmosphere altogether. They would one day command wingless machines with computerized controls designed to move through regions where there was no air, no up or down, machines that would fall like a manhole cover in the Earth's atmosphere. Though Yeager commanded and taught these aspiring astronauts the fundamentals of experimental flight, he himself didn't want anything to do with the space program directly. The traditional culture at Edwards was a most romantic and masculine one centered around flying faster and higher, drinking more and longer, and carousing; being the best by getting the job done right no matter what—and of course, surviving. The myth that astronauts were little more than trained chimpanzees had swept through the base a few years before, but by the time I arrived, all this was beginning to change.

The accommodations at Edwards matched the austere desert landscape. The ambiance of the place was most poignantly represented in the bachelors quarters where I stayed, which were more akin to barracks (with a communal toilet), and in the charred ruin of the mythic Poncho's Bar. In the evenings I taught classes of my own on energy optimization and space navigational theory to aspiring astronauts, and in the morning I breakfasted on coffee and donuts at the cafeteria before climbing into the cockpit of some jet to fly the day's mission.

Though I had come a long way since choosing this course of study, in a few weeks I would be 36 years old. My time had come. What I needed was that elusive phone call from a man by the name of Deke Slayton, who was head of flight crew operations down in Houston, and I needed it soon.

They knew everything about me—my education, my pilot skills, but most importantly, they knew of my ambition to join them in their mission. I had prepared myself in every way possible through flight and engineering training, and it had taken me 36 years to arrive at this level of experience. I had gained accolades and early promotions. I had a wonderful family. For the last nine years I had identified what I wanted in this life—but I needed that phone call.

Even in my boyhood on the plains of West Texas I could recall this impalpable sense that something extraordinary was just around the corner, without knowing exactly what it was. Now I knew. This was a voyage to that luminous crater-marked world that orbited our own, and I fervently believed that the machinery was now in place that could finally take me, or someone, there. Through the years I felt I had managed to position myself rather precisely at a critical juncture in the history of humankind. The phone, however, wouldn't ring.

Not until one evening in the late spring of 1966. Not until then, as I sat at the dinner table surrounded by my family in Los Angeles, would I know. The voice, a little fuzzy through the long-distance connection, identified itself as that of Deke Slayton. He promptly stated that he would like me to come work for him down in Houston, and I promptly said I would. The conversation was concise, with the air of firm decision about it. When I carefully placed the receiver into the cradle, I quietly contemplated the news. I then walked over and told Louise. We smothered each other in hugs, and took the girls up in our arms. Then we began our preparations for yet another life in Texas, where I would henceforth prepare for a very long, and, to me, sacred journey.

Sea of Sky

Preparation

4

The move to Houston was of a different sort than those before. This time there was a greater sense of expectancy, and I was hopeful that my own buoyed spirits would serve to make this transition less tedious for the family. Nevertheless, Elizabeth was concerned that the new owners of our old car might not take good care of it; Karlyn, now almost 13, was worried about how her tropical fish would endure the trip . . .

On one of my shuttles between Los Angeles and Houston, I drove out to the airfield with the jar of exotic fish in the seat beside me. Then I took the brightly colored creatures, suspended in their watery world, to the T38 trainer jet, and carefully strapped the bowl in the back seat. A few minutes later the fish and I gently arced into the pale blue sky over Southern California and headed toward that magical city of Houston just a few knots slower than the speed of sound. I'd been reminded that, a few years before, I had carried Louise's split-leaf philodendron to its new home in the bomb bay of a P2V while ferrying the plane to our new destination. I could do no less for Karlyn's fish.

We planned to build a new house in Houston near the space center, as the bankers now looked upon us more gently and nicely than before. Though we had always spent more on our quarters than the military housing allowance, the new place seemed to be a mansion. Life was good; a new momentum propelled daily life. There was the sense that we were being carried along by historic forces. The space missions occupied a portion of the nightly news, and what we saw on the television affected what occurred in our own kitchen. It was a strange and invigorating phenomenon. At the Cape, and here in Houston, the general mood of our astronaut group was manically gleeful—we couldn't believe we got paid for such work. Neither before nor after had I the privilege of working with such brilliant and motivated people as those at NASA in these years.

But the excitement of the new job and the hectic schedule of indoctrination, training, and technical duties was abruptly leavened by harsh reality before a year had passed. One afternoon in January as I stood in the concourse of LAX after a visit to North American Rockwell, maker of the Apollo command module, I was paged over the intercom. I knew no reason why anyone would need to get in touch with me at an airport. When I got to a phone it was Deke Slayton's office in Houston telling me of a fire in the command module at the Cape. Gus Grissom, Ed White, and Roger Chaffee were dead. A fire had swept through their capsule as it sat on top of the stack on the launch pad at Cape Kennedy. Three of America's astronauts had burned to death.

Forever after, this would be known as The Fire. But beyond the realm of personal loss, there was still a momentous task at hand. The price of exploration, each of us knew, was not paid in the coin of sweat alone, but in the ultimate currency of flesh and blood. The Fire took from us three of the planet's finest space explorers, but it also gave to us a much better insight as to just how daunting each detail of this trip to the moon would prove to be. And we were still on the ground.

There was a grim advantage to be had, because they did die on the pad, not in some silent sarcophagus hurtling away at thousands of miles per hour through the vacuum of space, forever unavailable to our engineers here on Earth. The men involved in the investigation that followed would make certain that Gus, Roger, and Ed did not die in vain. It would be transformed into a bitter blessing, a Pyrrhic victory that would one day quietly contribute to the success of the astronauts who followed. But for a time there was only grief. A few days after The Fire, Fred Haise, Ken Mattingly, Jerry Carr, and I would fly the four-plane missing angel formation over the memorial services in Houston, while our comrades were being laid to rest in the Arlington National Cemetery.

Because there was a political imperative involving the Soviet Union at the time, the public was unusually prepared to accept the loss of life in the interest of putting a man on the moon as soon as possible. This was the Space Race, the Cold War's more benign counterpart. Today the same atmosphere doesn't exist, and such a catastrophe—the *Challenger* explosion, for example—would likely shut down a project indefinitely, no matter how momentous. Even in the case of The Fire, there would be no launches for a year and a half, as its underlying causes were systematically corrected. For those who would follow in the footsteps of the *Apollo 1* astronauts, we grieved our loss, then responded by turning our attention with fresh vigor to each minute detail of the tasks for which we were given responsibility. Accidents could happen, errors would be made, failures in equipment would surely take place, because this was a program that's lifeblood was innovation, and even brilliant and dedicated humans miscalculate. Nobody had ever been to the moon. The combined task was to be flawless if possible, yet to have an escape hatch in place for when the unexpected presented itself. The degree of peril we faced was reduced to a number representing the probability that a series of critical parts would fail simultaneously. Numbers have a way of quelling fearful emotions, when we believe them. And they are hard to dispute even when you don't.

Our inner sensations were an eclectic mixture: the urge to climb a mountain just because it is there, and the simple desire to leave the world a better and broader place. In the face of our loss, it was still possible for me to imagine what Admiral Byrd sought while preparing to explore the sea of ice smothering the earth's poles, or what Columbus had in mind while charting what had yet to be charted. But more importantly, I began to see the larger project as but a sequence in the natural progression of civilization, which would eventually take us into the cosmos at large. NASA and its individual members were a part of the evolutionary process of humankind, a significant cog in the machine that would take us to other worlds. The people around me would be the first of our species to explore these new landscapes; Ed, Roger, and Gus were a part of this process as well, even in death. Though the idea was conceived for political advantage, going to the moon seemed fundamental in making the next step in discovering more about ourselves and the universe. We would not only take the first steps in this process, but also learn more about from whence we came by studying the mysterious world that had orbited our own for billions of years.

What I was interested in, and what everyone at some point in his or her life wonders about, is the origin of our existence, the grand question of how we came to be. We want to know if our existence has any larger purpose, and if so, just what it is. Human beings have pondered these questions ever since they were capable of any sort of meaningful contemplation. Ultimately, this was the impetus behind the Mercury, Gemini, and Apollo missions. These programs existed because Americans thought they might yield answers, and they were willing to pay millions of dollars to find out.

Irrespective of the individual motivation to be a part of the Apollo Program, each of its members were irresistibly drawn to the extraordinary adventure. It was understood that our three compatriots would admonish us, were we to abandon the larger project as a result of their death, because from the beginning it seemed unlikely that a project of this magnitude could be accomplished without loss of life.

When our distant progeny look back and contemplate the history of all living matter, they will view our time, with respect to the Apollo missions in particular, as the third major transition in the evolution of living beings. The first transition could be defined as that moment when sea creatures finally left the primordial sea and moved about on dry land; the second, when creatures took to the Earth's atmosphere; the third stage will likely be defined as that time during which we left the planet altogether in the latter half of the 20th century.

From an anthropological point of view, space exploration must be undertaken not only out of simple human curiosity, but also to further ensure the survival of the species. The 20th century has seen the unprecedented development and proliferation of magnificent technologies—many of them, through design, ignorance, or misuse, are capable of destroying life as well as enhancing it. Space exploration alone holds the promise of eventual escape from a dying planet, provided we wisely manage our resources in the meantime, and actually survive that long.

One of my first major decisions upon arriving in Houston was choosing to work on the lunar module itself as my technical assignment. Every week I would fly a T38 up to Bethpage, Long Island, to assist in the design, manufacture, and testing of the lunar module at Grumman Aircraft Corporation with Fred Haise. I recall the shock I felt the first time I saw a prototype of the machine that would one day land on and take off from the moon. I remember asking myself if we were really going to do it in *that*. It appeared so clumsy, top-heavy; so foreign to the act of flight. The insect-like posture seemed otherworldly. But in spite of its unconventional shape, it would prove to be an extraordinary flying machine. No one knew what a lunar module was supposed to look like, much less how it would respond to the pilot's commands. But this was an attempt by some of the finest technical minds in aeronautics and astronautics to fill the bill. Being the first manned craft designed to fly only in orbit, it had no aerodynamic surfaces, and in fact would never return to the atmosphere of its home planet. After making its rendezvous with the command module and safely delivering its pilots, each lunar module was scheduled to be retargeted toward the moon, and to crash into the crater-riddled surface. Its impact would be measured by seismic equipment earlier set in place by astronauts during their brief stay there. Innovation was a part of the daily routine, as there was no tried-and-true blueprint for what we were doing.

During the first Apollo missions I met the German successor to my old neighbor Robert Goddard, Wernher von Braun, in Huntsville, Alabama. This was the man whom I had heard so much about—much of it critical of his past. But the man I came to know was very different from the mythic reputation that preceded him. As it turned out, von Braun was a peaceful-minded man intimately concerned with the manner in which we would explore the cosmos, and he was leading the team that would build the launch machinery to take us there. He was also singularly passionate about his work. I came to learn that much of the lore surrounding von Braun was merely misinformed gossip. Many saw him as a thinly disguised and secretly unrepentant Nazi employed by the U.S. government, not the brilliant engineer who, as a young man, had his talents exploited by a totalitarian regime.

His life was larger, more dramatic than I thought possible. It was Wagnerian, operatic. While still a teenager, von Braun began his career in rocketry building the German Repulsor rocket for the legendary Hermann Oberth. As Hitler and his Brown Shirts came to power, von Braun's small group of engineers found financial backing in the Third Reich for their wild dream of one day sending men into outer space to explore the cosmos. During the war years, their designs for the A-1 and A-2 rockets (renamed the V-I and V-2 by Hitler himself) evolved light-years ahead of their counterparts in the Soviet Union and the United States. American designs were under the auspices of the brilliant but reclusive Robert Goddard in New Mexico.) The V-1 and V-2 rockets weren't effective militarily, as they couldn't carry a large payload of explosives, and weren't especially accurate. Moreover, production was peculiarly slow, which drew the suspicion of the military men. Finally, on March 15, 1944, von Braun was arrested by the S.S. and accused of sabotaging the rocket program. When asked how he had sabotaged it, the S.S. officer claimed that von Braun had been building rockets for the purpose of space travel all along, not war-making. But von Braun's boss, Walter Dornberger, insisted that such work did not constitute sabotage, and gradually the S.S. gave up its case against the young engineer. All of this was eventually neither here nor there, as von Braun and his team began actively—but secretly—seeking out the advancing American forces, rather than surrender to the Soviets coming from the East.

When they finally achieved their objective, just before the close of the war, von Braun was sent to the States, to White Sands Proving Grounds in New Mexico, not far from my hometown of Roswell. In the years succeeding the war, he expressed his wish to build his huge A-9 and A-10 rockets, which had been on his personal drawing board for years. He believed these immense rockets would be capable of sending large payloads, and men, into space. But, of course, the U.S. military wasn't interested, just as his former military bosses were not. What they wanted was a marriage between the two great military technologies that had come out of World War II—the rocket engine and the atomic bomb. They wanted ICBMs for a cold war. Space exploration would have to wait.

I saw the way he worked as much in the capacity of a humanitarian as a scientist. His work in building the Redstone and Saturn rockets was grounded to his belief that man himself must explore the cosmos, that we cannot be content in sending android surrogates in our place. From the dialog I began with him in Alabama, I came to understand the general nature of his mind. He was not only brilliant in the science of rocketry, but also extraordinarily prescient in his philosophy concerning realms of study that had yet to be seriously considered by mainstream scientists. This was, in a sense, the story of his life (in time, we would both discover we had much in common), but in an organization as large as NASA, bureaucracy was bound to wreak silent havoc and gradually impair the original vision. Men such as von Braun, I would soon learn, suffered real despair as the rigidity of mounting bureaucracy stifled innovation. Von Braun was a man who needed adequate terrain on which to implement such a vast dream.

A momentous week for me occurred in late summer 1971, after my spaceflight, and several months before von Braun left NASA for private industry. A NASA planning conference was held at a remote ocean resort, and I was the designated astronaut representative. Attendees were quartered in picturesque seaside bungalows. My housemates for this occasion were Wernher von Braun and the equally legendary Arthur C. Clarke, both world-class visionaries. The opportunity to privately probe the minds of these extraordinary men for several days still ranks in my mind as a highlight in my career, second only to spaceflight itself.

During the early days of the space program, a profound debate surfaced in scientific circles, almost from the moment Alan Shepard climbed into *Freedom 7* atop one of von Braun's tiny Redstone rockets and roared into a suborbital trajectory for the first time. The debate centered on this question of whether we should send men into space, or machines in our place. The argument continues to this day, as it has been renewed with each budget cut and tragic mishap. Taking men and women into such an unforgiving environment is both dangerous and expensive, but machines are of course expendable and far less costly.

Just after The Fire in 1967, the debate rose to a fever pitch, and in the 1970s, with the success of the unmanned *Voyager* projects, the question was raised again and again. It seemed to resurface with the transmission of each byte of data, and the reception here on Earth of each fantastically surreal photograph of the distant planets it neared. Finally, in 1986, with the fiery death of an entire shuttle crew, the debate seemed to come to a head. Why risk the lives of our bravest and brightest when machines can perform essentially the same tasks nearly as well, and much less expensively?

Oftentimes, the seriousness of the points in debate only thinly disguised the fact that allocation of funds was the underlying issue. Unmanned advocates felt shorted both in resources and accolades, ignoring the political reality that it was the public's personal identification with human adventure that made it possible to raise funds at all.

The public was launched into outer space through the experiences of the astronaut. Through the astronauts, the public walked on other worlds. Were we to ask any of the men and women in the space program who have died in the line of duty, or who will yet surely do so, I believe they would offer simple answers: We should go forward for reasons that nourish the human spirit, simply because we are capable of doing so.

My own thinking on human space exploration has deepened with the years, as research on intelligent machines and on the nature of consciousness has progressed. The differing viewpoints of astronauts, scientists, and others (funding issues aside) originate fundamentally from widely differing beliefs about the origin, meaning and destiny of life. Ultimately, our joint quest is to one day bring these questions to meaningful conclusions.

During the era of the Apollo Program of lunar exploration, it was conventional wisdom in both scientific and theological circles that not only were we alone in this vast universe, but also that the speed-of-light limitation from Einstein's theory of special relativity forever doomed us as a species to this middle-aged solar system on the fringes of the Milky Way galaxy. It is quite intriguing that a handful of scientists, the most prominent of whom were Frank Drake and Carl Sagan, were able to find the funds and government support to establish a radio astronomy project called the Search for Extraterrestrial Intelligence (SETI). Its objective was, and still is, to search the heavens for some coherent electromagnetic signal that would reveal the presence of an alien intelligence. This was established in spite of almost unanimous belief in the academic world of the period that life was such a freak accident of nature that it surely could not have occurred more than once. I was intrigued by SETI, for I had long questioned the dogma of our aloneness—it just wasn't reasonable to me. Also, Sagan delivered a number of lectures to assist us in the preparation for lunar exploration. I admired his boldness on this ET matter.

In the background, barely discernable above the noise of daily events, there had been persistent and serious claims beginning in the 1940s, coupled with adamant government denials, that extraterrestrial craft have been visiting Earth. I was much too busy with preparations for Apollo to be distracted by this controversy. During my operational years as a naval officer and aviator, I was well aware of the rules and procedures for reporting unidentified aerial sightings. Although I never had occasion to test the system by making a UFO report, I had heard stories circulating in the flying fraternity of the harassment, ridicule, and other difficulties suffered by those who did claim such sightings.

Claims and denials have continued unabated for decades. At the time of this book, however, there has yet been no hard, undeniable, physical evidence available in the public domain to substantiate such claims. But the *circumstantial* evidence has continued to mount as unusual sightings and bizarre events are reported and investigated by credentialed professionals. A significant number of persons, many quite elderly, formerly in military intelligence and government service, have privately dropped intriguing pieces of their stories on the ears of serious investigators, in hopes of getting the true story revealed, and in hopes of achieving amnesty from tough, punitive secrecy oaths they were forced to sign. In addition, certain documents from the presidencies of Harry Truman and Dwight Eisenhower have surfaced that add to the controversy about ET visitation and government cover-up. In this intriguing environment of claims, counterclaims, and denial, one might easily conjecture that even SETI, although it enjoys a certain scientific respectability today not in evidence when originally funded, was initially advanced as a smoke-screen and diversion from the real activities taking place behind the scenes. Garnering funds and political support for an advanced science project is nearly impossible without strong concurrence in the scientific community, and there was no strong scientific support in the 1960s.

5

These were the days when computer technology was relatively new, and this, in part, was what set our program ahead of the Soviets' in space endeavors. NASA designs took advantage of new innovations in lightweight materials and microcircuitry. Obviously, the smaller and lighter you could make your equipment, the less fuel your rocket consumed. You could go farther, faster, and for longer periods of time. The Soviets, however, not having the benefit of these technologies, were burdened with designing and developing monstrous rocket engines capable of lifting immense payloads to accomplish the same tasks. But at the time this proved beyond the state-of-the art of rocketry, so the Soviets were unable to send (and then recover) manned payloads to the moon.

Computer technology was something I understood quite well, having been introduced to it during my post-graduate work a few years earlier. I knew the principles by which they operated, and understood the analogies scientists were beginning to make with the human brain— the early work in artificial intelligence. But what I was growing fascinated with was not computer technology alone, but rather the nature of consciousness itself. My burgeoning interest was only a pale vision of what was to come. There seemed to be an entire panoply of ideas that the project of landing a man on the moon was stirring into awareness. Buckminster Fuller once said that to understand the human condition, one must begin with the universe. But a mystic would say that to understand the universe, one must begin with the Self. Not for 15 years would I recognize that both answers are correct, and converge to the same point. At the time I only began to perceive a glimmer that perhaps this project was part of some grand progression in evolution by which we enhanced the process of self-discovery. I recognized in the diminutive circuitry of computers an analogy to how our own consciousness was evolving. The computer hardware was analogous to the brain, the software and firmware programs analogous to the mentality that used the machine. Though brain evolution takes place on cosmic timescales, computer simulations of both computer and mental processes were progressing so rapidly that it would startle any eye that was open to observe. Though we did not recognize it then, humankind was going into space primarily to discover itself.

In December of 1968, we took that first step into deep space when Frank Borman, Jim Lovell, and Bill Anders left Earth orbit for the first time in human history, and struck out for the moon with the intention of orbiting that luminous gray world and returning safely to Earth. The timing of that flight was hurried, as there was new intelligence suggesting that the Soviets were ready to send men to the moon as well. Nevertheless, this was the first human foray into deep space. Seven months later, after the two additional build-up flights of *Apollo 9* and *10*, Neil Armstrong and Buzz Aldrin both set foot on the lunar surface, with Michael Collins minding the store above them. In July of 1969, men were walking on another world for the first time in the history of our civilization—or any civilization, as far as we knew. When Armstrong announced to the world that the eagle had landed, it was tantamount to that unceremonious yet monumental flop of the first sea creature that beached itself on dry land, and lived. Through the course of a few hundred thousand millennia, creatures of all description would then roam the forest, plains, and deserts, because the sea from which they sprang was no longer suitable for their kind.

So it seemed, and so it is, that humankind was embarking on an adventure that would take them throughout the cosmos, just as those first amphibians had led living matter throughout the dryer parts of the home planet. More than a quarter of a million miles away, as the astronauts' images were being transmitted from that rocky, dead world, the inhabitants of Earth seemed to grasp the weight of this notion, either consciously or unconsciously, and the astronauts themselves sensed the audience through the immense and empty distance. Giving weight to this assertion is the fact that during the intervening years since the moon landings, the photographs of Earth taken from deep space continue to appear daily on the pages of Earthbound periodicals and television, to stir in each of us a deeper sense of the larger reality.

Soon after becoming America's first man in space on the historic 15-minute flight of *Freedom 7*, Al Shepard had been grounded due to the diagnosis of an inner-ear disorder that caused nausea, vomiting, and a sense of vertigo, known as Meniere's syndrome. For many sufferers of the disorder, the symptoms clear up on their own, but for most there was little to be done about the attacks of vertigo and impaired balance. Such a diagnosis would, of course, put any pilot or astronaut out of business.

But six years later, in 1969, after a risky and experimental operation, Al was seemingly cured, and began lobbying NASA management for a slot on a lunar flight. Gordon Cooper was retiring, and he, along with Don Eisle and myself, had been the backup crew for *Apollo 10*, which presumably placed us in line as the prime crew for *Apollo 13*. Al successfully campaigned to fill Cooper's assignment, and selected Stuart Roosa and myself as his team. Stu was already as knowledgeable about the command module as I was about the lunar module.

Deke Slayton and the management at Houston had approved the move, but their decision was overridden by the powers that be at NASA headquarters. They felt quite strongly that Al should have more time to train for the complexities of a lunar mission. It was decided that we exchange missions with Jim Lovell's *Apollo 14* crew, who had trained as backup for Armstrong, Aldrin, and Collins on *Apollo 11*, so that the overall flight schedule would not be affected. Of course, we were all disappointed that our lunar expedition would be postponed for at least six months. But the decisions were prudent ones, even though, to the bull-headed test pilot's ego, they seemed more a personal affront.

But blessings often arrive unannounced. Sometimes they come disguised as misfortune.

In April of 1970 another Apollo stack stood on the pad at Cape Kennedy, poised to land two more Americans on the moon. I wouldn't be one of them, but that would come in good time; all traces of envy were now consumed by the job at hand. That day we had another flawless launch, another flawless translunar injection burn, and three of my friends were apparently headed for a full-up mission. Both the crew on the ground and the crew in space had this space travel thing down. For three days I listened to the happy progress on the headset as I sat before the console at mission control in Houston. The brilliant telemetry on the monitor before me showed all systems working in lovely harmony. Then one of the astronauts reported a strange, muffled "boom" coming from the service module. At the very same instant I saw the system parameters lilt simultaneously before me on the monitor. Then the dials silently fell to minimum levels. Fuel, power, battery, oxygen—all seemed to have fluttered into the vacuum of space. Then came Jack Swigert's voice, saying with stoic control, "Houston, we have a problem."

When I looked about the control room everyone seemed to be checking first their wits, then the telemetry. No one had any idea what had happened. Perhaps nothing at all but a loss of telemetry. After checking and doublechecking the data, however, we all knew it was for real. A grisly tension floated across the room. This wasn't erroneous information flickering through the maze of electronics. No, the data was accurate, and something had gone terribly wrong.

Within a few minutes we had identified the problem as an explosion of some sort in the service module. We all knew just how serious this was. Jim Lovell, Jack Swigert, and Fred Haise might not make it back alive. This was a very real possibility. Originally, they were headed for a rugged region of the moon called Fra Mauro, but a lunar landing was now a lost dream. The most we could hope for was getting them home before they depleted something they couldn't live without.

The training you have absorbed as a test pilot through the years automatically locks you into a peculiar pattern of thought and behavior during emergencies. You lose track of time. All that is on your mind is a series of the simplest declarative statements and questions you can possibly put to yourself. The craft is in trouble. What is the first thing to be done? The second? The third? What will it take to solve this one miniscule problem? No hand-wringing, no traces of panic. Silent prayers intersperse cool, even mental activity. The world becomes but an intellectual exercise; a friend's life is not at stake. Or so it seems in the vacuum of time. As I watched the telemetry appear on the screen I knew Al, Stu, and I had been spared this horrible fate. The thought lingered in a shadowy apse at the back of my mind. What lay at the front for the next four days was how we were going to manipulate the machinery. Three lives depended on just how well and how quickly all of us on the ground team could troubleshoot and solve a thousand miniscule problems.

Perhaps Stu, Al, and I prayed a little more fervently, as we knew just how easily it could have been us up there on the verge of freezing or expiring from carbon dioxide poisoning in a mortally wounded machine, headed into deep space at thousands of miles per hour. The stack of metal would become their sarcophagus, unless we could transform the lunar module *Aquarius* into a combination of lifeboat and auxiliary engine, and also keep the command module, *Odyssey*, alive to perform the reentry maneuver into the Earth's atmosphere. The crew was 200,000 miles from home when the explosion occurred, and it was clear they would have to continue around the moon before returning. In space you cannot simply stop and turn around.

The spacecraft was not on what is called a "free return trajectory," which meant that a burn was required before going behind the moon in order to achieve a path that would properly intercept Earth three days later. However, it was evident that electrical power to control the SPS (Service Propulsion System) main engine had been damaged in the accident and couldn't be used safely, if at all. If the burn wasn't executed precisely (if the engines burned too long, not long enough, or not at all), then *Aquarius* and *Odyssey* would float silently and irretrievably into deep space, a ghost ship sailing the cosmic sea. America stood a likely chance of losing three more astronauts.

However, the computers and descent engine of *Aquarius*—reprogrammed from landing on the moon to the emergency role of lifeboat—executed the burn flawlessly. As they passed around the far side of the moon, rushing through the blackness, they couldn't communicate with Houston for the cold, stony mass standing in the way. So for a while at mission control, and all around the world, there was only a deep silence coming from that lonesome craft in space, where earlier there had been the lively but weary voices of our colleagues. When they rounded the moon, their path bent by the lunar gravity, they should once again be able to communicate with mission control. Until then, each of us would wonder if there would be optimistic voices penetrating the silent darkness.

An hour later, the circuit came alive. All had gone well, and they were headed toward home. It had gone so well that the near-perfect burn required only a modest tweak the next day to improve the trajectory. What was needed now was a Herculean effort to stretch the life-sustaining necessities, such as water, fuel, and battery power. If this could be accomplished, the men stood a reasonable chance of making it at least within the vicinity of Earth alive. There seemed to be just enough of everything needed for a horribly cold three-day trip home, though no one knew exactly how to accomplish the task. The improvisation to make it possible would have to be downright ingenious.

All the while, word that *Apollo 13* was in danger had spread throughout the world, from Tokyo to Detroit to Tel Aviv. The world was hoping and praying—and watching. During the vigil, I spent my time, except for a few hours to sleep, in the lunar module simulator pretesting the protocol for the burns and devising procedures for manual control of the engines, should any subsequent failure of the lunar module computers take place. Ken Mattingly worked next door in the command module simulator conducting similar tests.

For four days the crew at mission control and the crew in the *Apollo 13* spacecraft worked through all the minutia in order to figure out just what it would take to get the three men home after such an explosion. Then on April 17, 1970, after nearly six days of traveling through the black void of space and bypassing their destination, the crew floated into the command module and prepared to make reentry into the Earth's atmosphere, not entirely certain that the electronic brain of *Odyssey* would "wake up" after her cold sleep, or that the parachutes would deploy properly. That day I left the simulator and returned to a position in the control room. My job was essentially done. And *Aquarius* had performed the lifeboat function flawlessly.

After several anxious minutes of communications blackout during the fiery ride to the sea, their voices came through to mission control, and moments later we could see that the three red-and-white parachutes had inflated beautifully on the huge monitor in the control room. Tears came to the eyes of grown men as they watched the end of the drama unfolding. The astronauts were safe and reasonably sound.

I recall emerging from my fatigue as though from a trance. The world around me suddenly came to life. A few minutes later we could see on the screen a helicopter hovering over the capsule as it deployed swimmers into the ocean. Eventually the hatch to the capsule would open in the midst of the South Pacific. Then three grateful, if gravely disappointed, explorers would emerge.

6

In the aftermath of the Apollo 13 mission, the machinery at NASA was activated to identify the cause of the explosion, and correct it. Apollo 14 had to be a flawless mission; the American public wouldn't tolerate anymore mishaps. We had already been to the moon. From here on out we had to make it look easy if NASA was to continue to garner funds from Congress.

While the task of correcting the cause of the explosion was undertaken, Al, Stu, and I continued preparations for the flight that loomed just nine months away. Fra Mauro, the rugged region of the moon where *Apollo 13* had originally planned to land, now became the scheduled landing site of *Apollo 14*. Intellectually we tried to distance ourselves from the similarities between our mission and theirs, but there was no dismissing the harsh realities of space exploration. What stood most prominently in our minds was the task of reorienting our preparation for the firsts that *Apollo 13* had been scheduled to accomplish. Each flight used the experience of previous missions and then extended the envelope of knowledge just a little further. Ours would be the first lunar landing mission to focus on science rather than establishing the techniques of space travel. However, we were scheduled to demonstrate a direct ascent to the command module upon leaving the surface of the moon, rather than the slower circling procedure. Instead of "chasing" the command module for one lunar orbit, the lunar module (LM) could feasibly fly straight toward it in one fell upward swoop. This, of course, would be done in the interests of saving precious time, which eventually translated into a smaller load of fuel and other consumables.

Al's project was especially daunting. The task of learning two new spacecraft from the ground up, along with our specific mission operations, within a period of 18 months, was awesome, but he was both a capable student and commander.

Each day presented itself as a miracle in these heady times.

I was finally going to the moon.

But in spite of the excitement and the demanding schedule, there were other matters weighing on my mind. In the weeks leading up to the launch date, I felt an unexpected upwelling of interest in those intellectual and philosophical ideas that had been with me since my evenings with Jim Remington in Massachusetts. The central questions I posed to myself concerned the relationship between the physical and spiritual sides of being, and how the two aspects came together in this vast cosmos we were setting out to explore. The lunar missions themselves had to be more than a stunt, more than a political dare. What I wanted to know were the underlying reasons we should want to go there at all. Why did it fire the imagination? What was it in our nature that compelled us to place so much at stake? Such questions, I thought, were the largest that could be asked. They all boiled down to whether we would discover our Earthbound viewpoints to be applicable in this brave new age, or simply parochial myth. I felt that such inquiry was, after all, the essence of exploration. Humankind's urge to journey into the cosmos was fundamentally grounded to our attempt to understand the cosmos in a concrete manner. We wanted to go there and learn firsthand how this universe came to be and what its underlying structure was.

The impetus behind my own curiosity seemed clear; my life thus far had been in part an attempt to resolve the paradoxical relationship between science and religion—perhaps because of my traditional and somewhat religious upbringing, and my adulthood, which I'd devoted primarily to intellectual and scientific pursuits. The standard models of reality with which I'd grown up began to appear so small—too small to contain this seeming dichotomy. Heaven wasn't likely to be somewhere just beyond Earth's atmosphere, as suggested by traditional thought.

A story told to me years later by a Russian colleague illustrates that the dilemma was not mine alone. It seems that when Yuri Gagarin, the first Soviet cosmonaut, returned from space, Khrushchev admonished him that though he likely had seen God in space, it was better they tell the people he hadn't. And so the story goes that when, on subsequent travels around the world, Gagarin encountered the Pope in Rome, the Pontiff requested that, though Gagarin probably had not seen God in space, would it be okay to tell the people he had? This story was accompanied by the uproarious laughter that only the earthy humor of a Russian can evoke.

My interest in resolving this issue took an unexpected twist at this time. Through the prompting of friends I studied Joseph Banks Rhine's investigations at Duke University into so-called paranormal phenomena, and quickly discovered that the literature was full of reports from a century-long investigation by a series of eminent scholars into these strange events. Being steeped in the traditional dogma of mainstream science, I often felt ill at ease as I read the accounts; the experimental results hardly seemed credible. But a thread of curiosity prevailed, though the scientific tradition I was so much a part of maintained that such events were silly, superstitious nonsense. The theology I was acquainted with claimed that the events were real, but either diabolically or divinely orchestrated—not of this world. It was in this realm of phenomena that the spiritual and physical realities seemed to meet. This was where the interface occurred. If it were true that supernatural phenomena *did* occur, then the scientific implications could not be more profound. If they did *not* occur, then the universe was far less mysterious than I had come to believe, and delusion far more prevalent.

Yet, as far as I could tell, a handful of scientists notwithstanding, there was little or no serious modern scientific investigation into the validity of such reported phenomena. Most scientists maintained that the fundamental issues concerning how the universe was structured were near to being solved. Complete understanding of basic principle, they believed, was merely a matter of time. The risk any serious investigator took in displaying interest in such fields was that of ridicule, and exile into scientific oblivion. But the ubiquitous beliefs resident in all the world's cultures, with regard to mysterious spiritual events, were for me not so easily set aside as superstitious nonsense. Two dominant bodies of thought (science and mysticism) in such sharp conflict was a situation I couldn't ignore, particularly as the issue seemed relevant to larger questions.

In the months prior to the *Apollo 14* launch, the crew would hide for a day or two every now and then for a rest. On one of these occasions I

From Outer Space to Inner Space

flew with a friend in his private plane to the Bahamas to scuba dive, and while there I met a medical doctor by the name of Edward Boyle, who conducted research at the Miami Heart Institute. Our meeting could not have been more timely.

One evening at dinner he introduced me to a friend of his, Dr. Edward Maxey, another physician, who was also interested in the full spectrum of consciousness. Dr. Boyle had been heavily involved in experimentation with hyperbaric chambers in the use of supersaturated, high-pressure oxygen environments that aided in the recovery of scuba divers suffering from the bends. Through the course of several visits, we discussed our private dismay with the scientific community, in their phobic reluctance to study paranormal phenomena. In the company of these physicians I felt confident I was talking with well-read, experienced, and competent men of science. There was the sense that something could possibly be accomplished with them, some new realm of the mind revealed.

And there was.

Out of our discussions, an idea came to mind, something we could do on our own that might shed light on our curiosity in these neglected fields of study. It occurred to us that there was an unusual opportunity presenting itself here involving telepathy. In the past, parapsychologists had found that the distance between two participants in any psychic experiment had no effect on the results. Moreover, the time between the transmission and reception of telepathic communications appeared to be instantaneous. What Dr. Maxey, Dr. Boyle, and I wanted to know was if results could be obtained over distances further than could be measured on Earth. Here was a rare opportunity to find out from a distance of more than 200,000 miles, 20 times as far as any two human beings had ever been from each other, the previous Apollo missions notwithstanding.

But being a personal, ad hoc experiment, we would have to keep the plans very confidential. It was innocent enough, but the press corps that blanketed the Cape searching for tidbits would blow our efforts out of all proportion were they to know, and certainly the NASA administrators would frown on any deviation from approved procedures on a moon-shot. But we primarily wanted to keep the project to ourselves in order to satisfy the personal curiosity we had in common. If anything interesting resulted, it could be published appropriately at some future time.

When I returned to Cape Canaveral after each discussion among the three Eds, as we referred to ourselves, I prepared for the mission that now loomed in the immediate future, my mind fortified for an onslaught of new detail. The attention at NASA was now focused on Al, Stu, and myself, and our preparation for one of the most audacious voyages ever undertaken. After landing on the moon, Al and I were to travel on foot more than a mile from the lunar module and climb to the summit of a meteor impact crater, collecting and documenting lunar samples along the way.

In spite of what occurred on *Apollo 13*, I don't think any of us were really concerned about a life-threatening mishap. At least for myself, I was chiefly concerned with only one thing: making a thoughtless mistake that might jeopardize the mission or embarrass the crew. The survival aspect of spaceflight is something an astronaut deals with very early in his or her career. Forever after your goals become increasingly narrow in focus: you strive single-mindedly to be as prepared as possible for each task you'll face during the course of the ultimate voyage.

As the launch date approached, there was a growing sense of confidence and unity with the crew and the extended team in Florida and Houston. We seemed to function with common mind, as if each individual had been absorbed into the anatomy of a larger organism. About a week before launch we seemed to have reached our peak of readiness, not unlike an athletic team that's on a roll and cannot lose.

Tense preparations, however, often gave way to surreal amusement. The day before launch, we were in quarantine to shield ourselves against the world's germs like futuristic prisoners behind glass, when our families and dignitaries arrived to wish us bon voyage. Kirk Douglas and Dr. Henry Kissinger were just departing when my family arrived. Louise, in her gracious and charming manner, conveyed a warm goodbye through the glass, and never betrayed the apprehension she surely felt. Karlyn, now a mature and precocious 17-year-old, alighted from the limousine and encountered Kirk Douglas. She introduced herself, so I was told, shook hands with the actor, and proceeded toward Dr. Henry Kissinger, to briefly engage in an exchange about world affairs.[1] Elizabeth, following her sister's lead, greeted Douglas, and dramatically curtsied, but added, "I didn't like you in Spartacus."

Then came the morning of January 31, 1971. The stack of *Apollo 14* stood white and brilliant against the black Florida horizon at the apex of lamplight beneath a new moon. To view it was to sense something alive in its presence as the systems were activated and came to operating speed. So much preparation by so many minds, so much careful attention paid for so long, and now here it was, the physical result of all those extraordinary efforts. This was the culmination of the history of man, the furthest mark made before he passed into that void beyond death. An Apollo rocket at night, flooded in white light, poised for a launch that would take it to another world and back, was a sight that stirred the blood.

I awoke on the morning of launch to a traditional breakfast with Deke Slayton; Tom Stafford, head of the astronaut office; Dr. Robert Gilruth, our beloved Houston boss; Al; Stu; and our backup crew of Gene Cernan, Ron Evans, and Joe Engle. The mood was jovial, notably lighthearted. We all knew what had to be done, down to the finest detail, and here we were about to make the voyage. It seemed as though the thrill was so great that it had to be contained, squelched in the interest of concentration. So we joked to break the tension; perhaps it was unseemly to be too lighthearted on such a heady occasion. After breakfast we gave our urine samples, were examined by the NASA doctors, then began the suit-up process. The countdown was well underway at the launch complex, and suddenly there was the sense that this was really it, we were on our way.

Once we were suited up and had finished the pre-breathing process, the ingress procedure was initiated. We took elevators down to the transfer vans, where an excited throng of family, dignitaries, and coworkers lined the walkway. We were three men who would soon be extraterrestrials, three men who looked and felt the part. The clamor of the applause was muted by the suit and helmet, producing the surreal sensation of doing publicly what had been rehearsed so often in private. As we reached the gantry and rode the elevator up, the sense of the stack's aliveness suddenly struck me. It was no longer a machine, but a living extension of everyone involved in this mammoth project. When we came to the swing arm leading to the open hatch, Guenter Wendt and his crew welcomed us and assisted our entry into the spacecraft, as they had for every launch since Al's first flight. Then we became part of its larger anatomy. Our unity was now whole.

We had arrived at the riskiest part of spaceflight. In the tedious hours of countdown all systems are painstakingly rechecked for their state of readiness. Just as the entire process was coming together, the violet shade of a thunderstorm appeared on the horizon. The radio in the headset came alive from time to time with local forecasts and personal predictions. Finally the squall neared, and rain slapped against the flanks of the rocket while we waited in cramped stillness. Time passed in tiny increments, but the clouds soon parted around the stack and eased over the sea. A thin new moon arose as though awaiting our arrival. Following the 40-minute lull in activity, I heard Al's voice in the headset call to rouse me; I had fallen asleep during the wait. We were go for launch.

Again, there was a brief final check to make certain we knew precisely where we stood now that the storm had passed. The countdown resumed; the numbers tumbled toward zero. This time there was renewed tension, the air about this familiar environment possessed an unbearable tautness. I could hear my heart laboring against the hectic activity of electronic voices in the headset. Upon this stack of metal and fuel sat a tiny pod containing human cargo. Soon the machine would come alive with a kinetic violence unlike anything any of us had ever experienced. A million gallons of fuel would ignite. A lurch, a violent vibration, a rush. Then three human beings would leave this world.

Yet these were thoughts and emotions that had to be suppressed as launch communications kept cadence for the final bit of countdown. As the rockets ignited, there was a gentle rumble. Then the seats began to shake as the engines brought their immense fury to bear. There was no mistake to be made; we were three men leaving the home planet.

Into the Vacuum

The Mission

7

In three days we would visit another world. The weight of that thought coincided with the enveloping rumble, more felt than heard, and the ignition of the rocket fuel as the stack slowly lifted from the pad. The sense that we were moving skyward pushed at us as the g-force accumulated, confirmed by the instrument panel.

Once the escape tower blew away, uncovering the windows, the eastern seaboard, the Caribbean, and the gentle bulge of the Carolinas were revealed. The horizon gradually drooped at the ends as the actual curve of the Earth became perceptible. After the first stage of the launch vehicle burned out, was jettisoned, and tumbled earthward, the dark line of the terminator, the division between day and night, rushed over to obscure the view and plunged us into darkness—a swift dusk brought about by our rapid movement toward the east. We were being hurled into space as though by a gigantic catapult until we had accelerated to the relative safety of a 100-mile orbit. All the white there were thousands of instruments and technical details to be attended to. Only when we were well established in orbit with system checks complete was there a moment to relax, to allow the pulse and blood pressure to soar, then settle back to normal. Then we could remove the helmet and gloves of the pressure suit.

Within only a few minutes, we had arrived. We were in outer space, that vast domain where I had once been taught the kingdom of heaven lay. Though space is only a vacuum, it is just as beautiful and strange as anything possibly conjured by a child's potent imagination. There is a sense of unreality here, with the absence of gravity and the tapestry of blackness broken only by an overwhelming glitter of stars that surrounded our craft. It occurred to me that the sky is not simply above the Earth—it is below and all around us, invisibly shrouding the home planet 24 hours a day out to the edges of the universe. At first Earth dominated our field of view, but slowly, gradually, as we sped toward that pristine bone-white world ahead that was yet but a little sphere, I saw how the planet is only one of countless celestial bodies. Yet the intricate beauty of Earth overwhelmed the senses. It drew the eye, commanded unanimous attention during stolen glimpses out the window. Until something went wrong.

Not three hours into the mission, just minutes after a flawless translunar injection burn that boosted us to escape velocity of 36,600 feet per second, we abruptly learned that our chances for a lunar landing were in jeopardy. Stu tried but could not dock the command module, *Kittyhawk*, with the lunar module, *Antares*, which still resided in the third stage of the *Saturn* launch vehicle. For some reason the docking mechanism was not functioning correctly. If this relatively routine maneuver couldn't be accomplished, then *Apollo 14* would be relegated to orbiting the moon, taking pictures, and heading home empty-handed after all the elaborate preparation for surface exploration. In the flight simulator, Stu had set himself the goal of docking with the smallest possible expenditure of fuel, but now that he was doing it for real, something wasn't allowing the LM to latch onto the nose of the command module.

On the first attempt, Stu hit *Antares* dead on, but the capture latches simply wouldn't set; on the second, he hit it dead on again, but the same thing happened. With each attempt we spent precious fuel needed for other planned operations on the trip. Stu backed us away a few feet and held position on the stack containing *Antares* while we discussed with Mission Control in Houston possible alternative procedures.

The problem, we figured, was with a piece of debris or frozen condensation in the mechanism deposited by the thunderstorm. It physically hurt to think that something as simple as this could cause us to abort a lunar landing. A dream blindsided by a detail. If we couldn't make the docking mechanism function automatically, then we needed to come up with plans to do it manually, perhaps even by suiting up and physically pulling the two ships together by hand as a last resort. Houston, however, wasn't hot for this idea. Even if it did work this time, what would we do the next time after making our foray to the surface and rejoining Stu in lunar orbit? Houston wasn't prepared to take the gamble of impromptu extravehicular activity in lunar orbit, though for the three of us, the risk wasn't an issue at this point.

About an hour after the first attempt, during which much contemplation and debate took place, Gene Cernan in Houston relayed to us another idea. Stu could try to dock again, but this time Al would flip a switch to retract the docking probe out of the way, allowing the final latches to engage. So Stu tried once more. Again, perfect alignment. Al hit the switch, and this time, after a halting silence, we heard the wonderful rippling racket of latches closing and joining the two spacecraft. For the time being we were back on a nominal mission to the moon. Here lives and missions were saved by improvisation.

For three days we flew across the abyss of space at the rate of several miles per second, making only a small adjustment along the way to assure our arrival at the precise point behind the moon to achieve lunar orbit. As the arbitrary calendar of days passed, the moon grew in the window, and the topography of the craters became clearly discernible. Slowly the view took on a surreal beauty.

Space travel requires intense concentration for long periods of time, interspersed with moments of high drama when either a burn is required or trouble arises. In the vocation of flying it is called hours-of-intense-boredom-punctuated-by-moments-of-stark-terror. However, spaceflight was—and still is—too new for actual boredom to arise. But it is not an exaggeration to say that with each burn the outcome is uncertain, and the success of the mission is at stake. Especially during the early space missions.

Before the construction of reusable shuttle craft, each piece of flight hardware was new and untested in the space environment. Unlike earthbound equipment, there is no opportunity for a test flight. And without precise execution of burns, the craft might assume a trajectory that would send it into deep space, skittering across the Earth's atmosphere, or into an unforgiving heavenly body such as the moon. Preparing for a burn is serious business, and before each one, Stu would announce, "It's sweaty palms time again, gentlemen."

But there was another task I'd planned to perform that wasn't on the checklist. And there were but four people on Earth aware of it. Every evening as the crew settled in for an attempt at sleep in zero-gravity, and the cabin grew quiet, I would take a moment and pull out my kneeboard, on which I had copied a table of random numbers, along with the five "Zener symbols" used in ESP experiments and made popular by Dr. J.B.

Rhine: a square, a circle, a star, a cross, and a wavy line. Then I promptly and discretely began the simple experiment that Ed Boyle, Ed Maxey, and I had devised. Not even Al or Stu knew what I was up to.

On four evenings, twice on our way to and from the moon, I matched one of the symbols with a random number between one and five, and then organized the numbers with a random number table copied from a math textbook. In this setting I would concentrate on a symbol for 15 seconds. Meanwhile, through tens of thousands of miles of empty space, my collaborators in Florida would attempt to jot down the symbols in the same sequence that I had arranged on my kneeboard. We each had a column of 25 spaces in which to write symbols for each of the six days I would be in transit. Then I zipped myself into my hammock each "night" after performing this seven-minute task. In the "morning" it was again business as usual, without a second thought of the experiment.

After three days, during which we traveled more than 240,000 miles and slowed into lunar orbit, Al and I climbed into the lunar lander, bid farewell to Stu, and prepared to undock. Stu deposited us in the 10-mile orbit from which we would descend to the surface, and returned through the vacuum to a 60-mile orbit for his work of mapping that barren world.

As Al and I orbited only a few thousand meters above the highest lunar peaks, the familiar gray landscape became recognizable. In a most austere way, it was hauntingly familiar. There were mountains and valleys, a sun in the sky. For the first time in three days there was a relative up and down; the tiny earth in the distance appeared a satellite of the moon. But little else resembled anything we had ever seen. No flora, no softening features. Here the land was presided over by an omnipresent night sky. Without an atmosphere, sunlight lent an unreal clarity to the landscape. If we were to open the hatch, it seemed as though we could reach out and run our fingers over the ragged lunar surface.

As for our the mission, there was a tentative sense of optimism. We knew we were over the right area because we could see Cone Crater and the entire topography of Fra Mauro beneath us coming into lunar sunrise, just as we had imagined it would appear out the window of *Antares*. Then I was struck with a sense of wonder. The same sun rose this morning over the Atlantic, I thought, as we silently flew over this strange world.

As we flew through the terminator into the lunar night, barely 90 minutes from beginning our powered descent to the surface, trouble arose once again. The computer guidance software, we discovered, was receiving incorrect information, thereby producing a signal that would abort the landing once it began. Moreover, we would soon be passing behind the moon into a communications blackout, cutting off our life-line of information and support from Houston. Once again, we had to carefully, accurately, and quickly find a way to bypass a problem that threatened the mission. But of the 90 minutes available to us to solve the problem in this case, we would be behind the moon and out of radio contact for 60.

In Houston, where it was 2 o'clock in the morning, they were working on the problem. When we emerged from behind the moon, we expected they might have procedures for a fix prepared for us, likely a revised computer update to get around the problem. But it was much more than that. After regaining communications we would have but a stark 10 minutes to complete the pre-descent checklist, enter navigation updates, and manually program the remedy into the computers. Because of fuel constraints, time was not on our side. We simply could not afford another trip around this rugged world without major changes in our cramped schedule.

By reprogramming the computer manually with Houston's instruc-tions, we could get it to ignore the false signal and not actually initiate an abort when the engine ignited. But there was a penalty in this. If an abort were actually required anytime during the landing, we would need to perform the lengthy series of tasks manually and fly for a time without computer assistance. In order to continue the mission, we were surrendering the help of an automatic abort system that could deliver us into a safe upward trajectory at the push of a single button, were another emergency to occur during or immediately after landing.

The entire team performed magnificently during the 90 minutes available before descent. All the required tasks were completed, with a few seconds to spare, as we rotated *Antares* into position for descent engine ignition.

As is often the case, however, errors in a system tend to propagate. Although no one realized it at that moment, what we had done to fix the abort switch would cause the computer not to recognize and lock on to the radar signals bouncing off the lunar surface as we descended and approached the landing site. Thus, we couldn't check our altitude with these updates, nor could we just look out the window, as we were quite literally on our backs, feet forward during powered descent, the windows displaying nothing but a striking pattern of stars. What made this particularly unsettling was that the Fra Mauro region happened to be a rather rugged area of the moon, filled with hills, valleys, and craters. Even if our landing approach was perfect, we wouldn't have the benefit of a computer abort system in case of trouble with the terrain. But more immediately, our mission rules forbade us to descend below known mountaintop levels without the radar—a height we would reach within the next minute or so.

We worked quickly, our eyes sweeping the control panel, hoping to spot the problem that prevented the radar system from measuring the surface of that meteor-scarred region into which we were rapidly descending. I recall the unusual sense of detachment, one I'd known before on occasion, in which the mind focuses impersonally on the pattern of required tasks. Feeling and emotion were vanquished, and just a body was left, automatically performing the job, searching for a solution to a dilemma, functioning as an extension of the computers at my fingertips. Then something occurred to me in this trance-like state. Through some deep recess of memory I realized that the radar might need to be reset after the abort switch fix, and that there were but two quick possibilities. But which to try first? Just then, as though reading my mind, Houston came in with the right call. Al recycled a circuit breaker, the radar locked on, and we could see that the data we were receiving was accurate and the computer had been guiding us perfectly. The mission had been salvaged once again with the help of Houston, this time within scant minutes of landing.

As we slowed our descent and pitched forward to a "feet down" position in the manner of a helicopter, the moon's rocky horizon rose up, its broken gray surface appearing in the fine texture that is perceptible only at close range. And there loomed Cone Crater, just as in simulator training. After locating what appeared to be a smooth landing zone, and easing *Antares* toward it, we could soon see the mysterious lunar dust being swept away, as the craft's descent engine was now close enough to stir it up. The landscape was altered. We were lowering into a silent, dead world, a place where nothing moved. All lay still, except for our craft, just as it had, seemingly, since the beginning of time. Eventually *Antares*'s landing sensors touched the soft lunar soil, and the engines were promptly stopped. We dropped the remaining inches to the surface. We had arrived. Alan Shepard's and Edgar Mitchell's craft was resting on the face of the moon. Two extraterrestrials in a very foreign land.

8

During such drama, relief and elation are momentary. They rise to the surface in soft, tingling flashes. After the ship's circuits were methodically safed and reconfigured for emergency liftoff, there was little more emotion shown than an ecstatic grin. The cabin fell quiet as we both listened for that elusive sound of safety. We were on the moon, secure for the moment, yet joy was eclipsed by alertness. There is a sense of wariness you guard with every bit of conscious candlepower at your disposal at such times. You recognize this faculty as one of the vital mechanisms that has kept you alive and well all these years, and you do not betray it by indulging in excessive celebration, even after landing on the moon. A handshake and a word or two would suffice for the moment. After taking careful accounting of the condition of our craft, performing our respective checklists, and exchanging status reports with Houston, we were ready to begin the procedures that would lead to opening the hatch. Then we would leave the sanctuary of *Antares*.

A rumor circulated widely for many years following the landing that I had questioned Al as to his intent had we arrived at the minimum altitude without landing radar. Would he have called for an abort, or continued? His presumed response was: "You'll never know!" I must report, that conversation never took place, as we both knew that the abort software problem had forced us into manual abort mode, not automatic mode. We would have been required to manually pitch forward as the first step of the abort, and there before us would have been the landing site, just as we had practiced. Computer guidance was dead-on even without the landing radar. Without question, we would have proceeded.

The plan of surface exploration called for two 4½–hour excursions outside, during which we would erect the flag, set up the television equipment and scientific instruments, collect lunar rock and soil samples, take hundreds of pictures, and then make a trek by foot up to the rim of Cone Crater. We believed the summit would provide a sweeping view of the ancient aftermath of a cataclysmic meteor impact that spewed layers of debris over the lunar surface a billion years ago. The diameter of the rim was more than a thousand feet across. It was this field of ejected debris from the crater that earthbound geologists wanted us to sample in order to find clues to the inner structure of the moon.

There is a dramatic difference between viewing a landscape from behind a window and walking out into it yourself. By entering it and walking among its hills and valleys you become a part of its topography, a part of its history. When Alan opened the door to the lunar module and descended the ladder to the dusty surface, with me following a few minutes later, I felt we were suddenly native to this land—the only ones it ever had. The stillness seemed to convey that the landscape itself had been patiently awaiting our arrival for millions of years.

Though there isn't such an emphatic sense of "down" on the moon due to the reduced gravity, there's no doubt that you are walking about the surface of another world—a stunningly beautiful and foreign world. The sheer eeriness of the view assaults the senses. The shapes and starkness of the sun-drenched landscape are more dramatic than similar geologic forms on Earth, which are softened by atmospheric diffusion. The glare of sunlight relentlessly burns at the edges of shadows, and there is the startling sense of silence in this land that has never known sound. Beyond the curvature of glass of my helmet, inches from my face, lay an infinite vacuum.

There was work to be done before we could begin our journey. During the first outing we would set up a thermonuclear station, which would power many of the scientific instruments for years to come, as well as the television station, which would transmit the progress of our journey to an enormous audience of Earthlings more than a quarter of a million miles away on the beautiful blue and white planet that loomed directly overhead in the black sky. Time was always draining away. Long checklists had to be attended to in order to assure that nothing would be missed. Our presence here had cost the American taxpayers millions of dollars, a fact that wasn't lost on either of us. Each minute had to count for something. Such thoughts came to mind as a recalcitrant fitting or a stiff fastener bled a few more seconds from the schedule. Meanwhile, we kept our wonder in check, or at least under our breath.

After our first "day" on the moon, we tried to sleep through the artificial night. The lunar day consists of 28 Earth days, so dusk is almost always a long way off. But after that first day of work, Al and I retreated to the lunar module, where we closed the blinds and crawled into our hammocks, which lay crosswise, one below the other. Then we attempted to sleep on the surface of the moon.

Though exhausted, rest would be at best fitful, as we had landed on a slight incline. Always there was the sensation that the lunar module was about to topple over, due to the slope and the lessened sense of "down." With only a sixth of normal gravitation tugging at the fabric of our bodies, there was the phantom feeling of instability. This led to the tendency of the imagination to run away with itself, though intellectually we knew what caused this sensation, and that we were quite stable. But if *Antares* did topple over, we would doubtless be stranded on the moon for the rest of our brief lives, with only a few hours of oxygen and other vital supplies at our disposal. Such thoughts produced a strange and subtle energy when it came time to rest. Edgy half-dreams would surface. So we more or less spent the night listening to the tiny sound of an occasional micrometeorite colliding with our fragile home, our minds secretly turning over the knowledge that we were the only two living creatures on this dead world. The knowledge that survival was not guaranteed mingled with the exhilaration of being the first to explore this place. Two extraterrestrials asleep in their spaceship.

When it was time to arise a few hours later, we knew that our day was centered around a trek to the rim of Cone Crater. We would command a view that no humans had ever beheld: an ancient lunar crater 750 feet deep and 1,100 feet across. A general fatigue from sleeplessness was surmounted by thrilling anticipation. Again, we pulled on our extravehicular equipment, depressurized the cabin of *Antares*, and walked into this strangely lit world where sunlight left black shadows. We then loaded equipment on the MET, or modular equipment transporter, consulted our checklist, and set out for the rim of Cone Crater, the summit of which could be seen on the eastern horizon.

We left *Antares* surrounded by a cluster of scientific equipment we had assembled and placed in position, taking with us our MET, the first wheeled vehicle on the moon. Stone Age technology somehow seemed fitting here: The MET was a sort of wheelbarrow to be carried backwards, a rickshaw with a single handle. As we set out for the summit, we would look back from time to time to see the silvery pair of tire trails leading all the way back to the spider legs of *Antares*. When looking toward the sun in the strange glare of lunar light, the tire tracks looked like the greasy trails of earthly slugs. The scene as a whole was so otherworldly, at once hauntingly familiar and unfamiliar.

Embarking on a journey on the moon by foot was a more puzzling experience than anyone had anticipated, certainly more difficult than could be imagined by merely studying high-resolution photographs of the surface, which were what we used to navigate by. Landmarks clearly depicted on the photographs were obscured by larger than expected undulations of the cratered surface. This was a terrific surprise. Though the area around the landing site was relatively smooth, the overlapping of ancient craters left ridges often 2 meters high that hid the navigation points. Micronavigation (knowing our location within a few meters), as desired by the geology team in Houston, proved to be impossible with the equipment at hand. Distances became plastic due to the nearness of the horizon on this small planet and the unreal clarity of the airless scene before us. Our estimates of distance were in error by 100 percent, as objects typically appeared to be at half the distance they really were. And there were other problems to be surmounted. The stiff pressurized suits fought each intricate movement as we conducted the delicate tasks of documenting and collecting samples and performing soil experiments. We were Earthlings in a world that possessed its own dimensions.

As we mounted the gentle slope and were climbing in earnest, we were more than twice as far from our LM as any of our predecessors had been, yet less than halfway to the summit, and several minutes behind our time line. The precious minutes lost in the futile search for landmarks and the exhausting climb caused our heart rates to climb also; because of this, consumables in the life-support system diminished at a faster rate. But things would work out—just a little more slowly than had been so rigidly planned. We took comfort in the fact that the work was getting done in good fashion, in spite of the obstacles. And there was comfort to be had in the sight of *Antares* squatting in the distance below us, ready to provide safe haven if needed.

As we slowly bounded along in the strange bouncing gait required on the lunar surface, strangely cheerful yet frustrated, I suddenly recognized a distinctive landmark I thought we'd already passed. With grim reluctance I told Fred Haise in Houston that all of our previously reported positions were in doubt. As our hearts raced and our lungs gave our blood precious oxygen, carried through 250,000 miles of space, the frustration grew. There was very little Houston could tell us, so we continued the climb, sometimes carrying the MET, as it bounced menacingly in the reduced gravity over the rough terrain. All the while, the summit of Cone Crater refused to reveal itself.

A few minutes later Fred's voice pierced the quiet, sounding grave through the vast emptiness. He was about to have us turn back, we knew, but from what we could tell from the map and what we saw above us, the summit of the rim couldn't be much further. We asked for a little more time, almost pleaded; what lay ahead was something neither Alan or I could surrender easily. This was what we had come here for. If we turned back now, it was lost to us forever. Houston seemed to understand this. A few more minutes of laborious breathing passed, and we received the happy news of a reprieve of 30 minutes to see if we could reach the rim. The flight surgeon in Houston merely requested that we take a minute to rest, as he saw our hearts racing on the monitor.

After a short pause we resumed the struggle up the slope, carrying the MET between us as you would a stretcher. Though it was tempting to leave the contraption behind, it carried valuable film, tools, and space for samples, all of which we would need when we reached the summit. Without the MET, the walking would be far easier, but we knew the Houston geologists were in a frenzy to get their hands on samples of the Volkswagen-size boulders that littered the rim. This was science. This was the mission's larger purpose.

As we continued the trek I was struck by an upwelling of obscure feelings. The sheer beauty seemed to summon some deep nocturnal emotion. I was longing for something, I realized—perhaps one of those precious moments when I might stand quiet and alone, gazing out over the terrain of an ancient world whose history is measured by a scale of billions, rather than hundreds, thousands, or even millions of years. What I craved was a moment to contemplate my own place on its face and the one that stood at our zenith. The view magically altered perspective, allowed for new vantage points. But Al and I always had to keep moving. There was work to be done, and a mountain of the moon to climb whose summit lay somewhere in the distance.

A few minutes into our reprieve, Fred Haise's voice again broke through the sound of heavy breathing, telling us it was time to stop, consider our current location as the summit, and perform the required tasks. Then we were to head back. This time it was for real. With dour frustration we did our work, our hearts pounding in our ears against the din of voices in the headset. Then we stole the briefest of moments to look out over the plains of Fra Mauro from this exalted height before orienting ourselves for the trek back toward *Antares*, visible on the barren plain below. Because we would not see the rim of the crater, I took a few seconds to allow the scene to make its impress on my memory.[1] Though we still had two more sample steps on the way back, from now on Al and I would be earthbound.

We kangaroo-hopped our way back down the slope, taking giant leaps through the void that is the lunar atmosphere. Only a faint gravity brought us down. Within the span of a few minutes we had performed the additional tasks, taken a moment for Al to make a renowned golf shot, and me a less-famous javelin throw. Then we climbed back into the *Antares*, stowed our samples, and prepared for the Great Elevator Ride to the command module where Stu hung somewhere in orbit. We finished our housekeeping while waiting for Stu to come to the right place overhead, then pushed the final computer key that sent us blasting upward. The dust and rock that scattered in our wake would remain just as it fell for millions of years to come. As we saw the surface of the moon diminish, both Al and I were struck with a strange nostalgia for this world. We would not be this way again.

9

After carefully timing and coordinating the moment of launch, Al and I flew directly from the lunar surface to the command module. As we silently approached the beautiful metallic spacecraft, there was a momentary resurrection of the old apprehension from the last time we tried to make a dock. Stu was our ride home. Was this going to require a spacewalk? Would we be able to dock at all? Such questions served to cushion against total surprise. But no sooner had we made the first attempt, than Stu had solidly captured our craft.

We were all three enormously relieved and happy at the successful capture, and allowed a bit of levity to show. As we prepared for transfer to the command module, and knocked twice on the hatch to signal readiness, Stu was unable to resist the temptation of the moment. He responded to the knocking with: "Who's there?"

In space there are nearly 10 times more stars visible to the naked eye than on Earth because there is no atmosphere. Likewise, familiar objects are approximately 10 times brighter. Stars and planets seem to burn against the cool blackness. There is the sense of being swaddled in the cosmos, surrounded by the beautiful silent glitter of the Milky Way and all the galaxies beyond. As we departed our lunar target and sailed home through the vast emptiness, we rotated in what is called the barbecue mode, slowly turning in order to sustain the same thermal effect on all sides of the craft. Earth imperceptibly grew larger with the passage of time. In the quiet hours just before our designated night, I would pull out the clipboard and perform the experiment with my friends in Florida. Then I would drift into a gauzy sleep. There was an impalpable sense of satisfaction, safety, and well-being—a sense I hadn't experienced for several days.

A wonderful quietness drifted into the cabin, the satisfying glow of a job well done. The lion's share of my own work was complete, and all I had to do was monitor the spacecraft systems, which were functioning perfectly. Now there was time to quietly contemplate the journey. I could lie back in weightlessness and watch the slow progress of the heavens through the module window. My mind ebbed into that quiet state I had longed for on our trek to the rim of Cone Crater. There was a vast tranquility, a growing sense of wonder as I looked out the window, but not a hint of what was about to happen.

Perhaps it was the disorienting, or reorienting, effect of a rotating environment, while the heavens and Earth tumbled alternately in and out of view in the small capsule window. Perhaps it was the air of safety and sanctuary after a two-day foray into an unforgiving environment. But I don't think so. The sensation was altogether foreign. Somehow I felt tuned into something much larger than myself, something much larger than the planet in the window. Something incomprehensibly big. Even today, the perceptions still baffle me.

Much of my thought and feeling at the time has since undergone a process of alchemy. Contemplation and the process of resurrecting memories has perhaps served to illuminate the shadows of such a peculiar event, but the tableau is so vivid as to have lost none of its clarity. It looms in my memory with extraordinary resolution. There was the initial awareness that the planet in the window harbored much strife and discord beneath the blue and white atmosphere, a peaceful and inviting appearance. On a small peninsula of Southeast Asia, a brutal civil war was being waged within the thin canopy of foliage. This was a war that commanded the attention of another country defined by invisible borders on the other side of the planet. I knew that my younger brother and his Air Force colleagues were flying their missions there. Then, looking beyond the Earth itself to the magnificence of the larger scene, there was a startling recognition that the nature of the universe was not as I had been taught. My understanding of the separate distinctness and the relative independence of movement of those cosmic bodies was shattered. There was an upwelling of fresh insight coupled with a feeling of ubiquitous harmony—a sense of interconnectedness with the celestial bodies surrounding our spacecraft. Particular scientific facts about stellar evolution took on new significance.

This wasn't a religious or otherworldly experience, though many have tried to cast similar events in that mold. Nor was it a totally new scientific understanding, of which I had suddenly become aware. It was just a pointer, a signpost showing the direction toward new viewpoints and greater understanding. The human being is part of a continuously evolving process, a more grand and intelligent process than classical science and the religious traditions have been able to correctly describe. I was part of a larger natural process than I'd previously understood, one that was all around me in this command module as it sped toward Earth through 240,000 miles of empty black space.

This new feeling was elusive, its full meaning somehow obscured, but its silent authority shook me to the very core. Here was something potent, something that could alter the course of a life. I wondered if Stu and Al were experiencing it as well, if they sensed the profoundness of this environment, but being busy with other tasks, they showed no outward signs. Somehow I never felt the urge to ask.

Billions of years ago, the molecules of my body, of Stu's and Al's bodies, of this spacecraft, of the world I had come from and was now returning to, were manufactured in the furnace of an ancient generation of stars like those surrounding us. This suddenly meant something different. It was now poignant, and personal, not just intellectual theorizing. Our presence here, outside the domain of the home planet, was not rooted in an accident of nature, nor the capricious political whim of a technological civilization. It was rather an extension of the same universal process that evolved our molecules. And what I felt was an extraordinary personal connectedness with it. I experienced what has been described as an ecstasy of unity. I not only saw the connectedness, I *felt* it and experienced it sentiently. I was overwhelmed with the sensation of physically and mentally extending out into the cosmos. The restraints and boundaries of flesh and bone fell away. I realized that this was a biological response of my brain attempting to reorganize and give meaning to information about the wonderful and awesome processes I was privileged to view from this vantage point. Though I am now more capable of articulating what I felt then, words somehow still fall short. I am convinced that it always has been and always will be a largely ineffable experience. What was clear, however, is that traditional answers to the questions "Who are we?" and "How did we get here?" as derived both by science and religious cosmologies, are incomplete, archaic, and flawed. There is more to the process than we have yet dreamed.

Three days after our return from Cone Crater we approached Earth's thin layer of life-giving atmosphere. The blunt ablative heat shield slammed into the upper reaches of air at 36,000 feet per second, slowing us with a force of 7 g's for a few seconds. After the anticipated communications blackout of reentry, during which the outside of the capsule reached temperatures in excess of 4,000° Fahrenheit, the red-and-white parachutes blossomed out above us. Suddenly we were Earthlings, back from our voyage. We had brought with us a precious cargo of film, rock, experimental data, and mental images of another world. And we had survived. After a few minutes of swinging by the pendulum of parachute fabric, the cooling heat shield of the capsule splashed into the Pacific Ocean within sight of the waiting recovery forces. As we bobbed about, we again took an accounting of ourselves and shut down our craft. We waited for the recovery swimmers to deploy the life rafts, open the hatch, and give us the biological masks that would protect the world from any mysterious virus that may have infected us on the moon. Then within but a few minutes of splashdown we were in a helicopter hovering over the gray deck of the USS *New Orleans*.

We arrived in Houston with our lunar treasure trove, which made the geologists at the Lunar Receiving Laboratory ecstatic. We had collected, as was the plan, more rock samples and data than was available from previous expeditions—nearly a hundred pounds in all. For three weeks we worked in quarantine with the geologists, doing what we could to assist in identifying where we had gathered particular samples, debriefing flight personnel, and of course writing reports.[1] When I finally had an opportunity, I telephoned Ed Maxey in order to collect the results of our private experiment. I was also in telephone contact with Dr. Rhine, who graciously offered to take all the original data sheets from the participants and run a statistical analysis of the experiment in his laboratory, even though there were only 150 data points—a rather small number compared to the thousands of runs he had conducted. He also suggested that another laboratory be used as an independent check.

We chose the assistance of Dr. Karlis Osis in New York, a well-known researcher in the field. We could see from simple preliminary analysis that the results were likely to prove quite interesting. Dr. Rhine suggested that if the procedures for conducting the experiment were without flaw (which they proved to be), the results should be published immediately, regardless of whether they were positive or negative. This was a seminal experiment in the space environment, which gave it special pertinence. He even offered to coauthor and publish a paper in the June issue of *The Journal of Parapsychology*. There were, however, forces gathering that, in the public eye, would taint the work we had done; forces that would place a sensational spin on the project, rather than a scientific one—or no spin at all. Such forces are difficult to anticipate.

In the last weeks before the moonshot, during the time Ed Maxey, Ed Boyle, and I were laying out the details for the experiment, Ed Maxey suggested we bring on board a man by the name of Olaf Johnson, who was a professional and reputedly competent psychic. This made a lot of sense. What we wanted was a mix of everyday individuals and experienced, competent psychics. So I agreed that Johnson could be added, believing that the private nature of our efforts would be respected. Unfortunately, Olaf found it difficult to contain himself shortly after my return, and leaked our project to reporters. The result was disastrous.

All across the country, in nearly all the major newspapers there was mention of clandestine psychic experiments conducted on the Apollo 14 moonshot. From Berlin to Beijing, sensational headlines shredded the project, the results of which they didn't know. And there was nothing I, or anyone, could do.

One morning during the first week of quarantine, Al and I were at the breakfast table when he came across an article in the stack of newspapers entitled, "Astronaut Does ESP Experiment on Moon Flight." He chuckled as he read, certain it was an absurdity dreamt up by some creative reporter, and leaned over the table to tell me about it. He then buckled with laughter. After an awkward silence, I told him that I had in fact done it. He looked up, nonplused. But a moment later I thought I saw another glint of laughter spark in his eye, as he silently returned his attention to his breakfast plate. The subject was never brought up again.

The results of the experiment were dramatic, but they had to be properly understood. When viewed through the prism of statistical probability, they were profound—therefore, more striking to the professional than the man on the street. In many ways they revealed the processes at work. But the press seemed interested only in the headline: Astronaut Does ESP Experiment on Moon Flight. Misinformation abounded. Somehow word got out that the experiment turned out to be negative or without significance, when in reality the results were completely in keeping with the thousands of experiments conducted both before and after our flight in laboratories around the world. It appeared that even great distances of hundreds of thousands of miles did not alter this mysterious means of communication.

The results of our in-flight effort suggested that there was some kind of communication being achieved during the experiment that wasn't through the conduit of conventional transmission. When we compared my four sets of data (two on days outbound to the moon, and two returning) with the six data sets of the individuals here on Earth, we saw we'd achieved what is known as a "psi missing" result for the days I actually did the experiment, and precisely "chance" results for the other days.[2] The psi missing statistics were such that there was but a 1-in-3,000 probability that the results were a random happening.

Psi missing is a well-documented occurrence in parapsychology normally associated with the state of one's belief. Dr. Gertrude Schmidler thoroughly investigated such events many years earlier, and coined the term the "sheep/goat effect"—sheep being those individuals who achieved positive scores well above chance results, and believed that they could; goats being individuals who didn't believe they could achieve *any* results, but do so nevertheless by scoring significantly *below* chance. If one were to guess at the results of a series of coin tosses, for example, and correctly guessed 75 out of 100, or missed 75 out of 100, then both would be equally significant statistically, and would suggest something about the state of each participant's state of belief during the process. What we could not control was the influence of cultural bias (about which many books have now been published), that was bred into the individual belief systems of those reporters and editors who would interpret for the public at large and publish their own opinions. In classical physics, the individual's belief system does not matter. However, in this realm of science, I would later discover that it not only comes into play, but it is also fundamental.

The NASA administration never chastised me for the project, and a large number of space center personnel even dropped by my office, furtively, to inquire about the experiment. But there was forever afterward an undercurrent of disregard for such studies by nearly all the administrators at NASA—all, that is, but Wernher von Braun. One day after leaving quarantine, he contacted me privately to tell me he understood what I wanted to do. He then hinted at the possibility of using some of NASA's resources so that we might be able to accomplish some of this work. What he wanted me to do was to conduct a survey and assemble a list of NASA facilities, people, and equipment that might be useful for some of the consciousness studies that consumed both our interests. Unfortunately, before my survey was well underway, von Braun left NASA for private industry, thoroughly discouraged that the budgets for space efforts were being reduced. There would be no lunar surface missions after Apollo 17, and certainly no mission to Mars. The American public had had their fill. Therefore, there would be no need for the construction of the huge rockets needed to take man throughout the solar system in the foreseeable future. So von Braun simply left.

After the quarantine, I had a six-month obligation with NASA headquarters to be on call for public relations work. There were visits to foreign capitals, the Paris Airshow, educational projects, and politicians who had enough budget clout with NASA to pry an astronaut loose for an appearance with his constituents. I had also agreed to back up the *Apollo 16* crew, knowing that this would be my last crew assignment with NASA unless I was willing to wait another decade for a shuttle to fly. Ground assignments within government simply held no appeal.

The atmosphere was tumultuous at NASA in 1971, as a bureaucratic inertia was setting in. Kennedy's mandate 10 years earlier had challenged the nation to put a man on the moon by the end of the decade. It was declared primarily for political reasons, and it wasn't an open-ended mandate for a series of missions that would initiate exploration of the solar system at large. The decade had drawn to a close, the project had come in on time and below budget. Now money was the big obstacle, and the American people decided, more or less by default, that the next epoch of exploration was not yet at hand.

Down and In

The New Journey

10

Since I was a boy, I've lived, in a way, two lives, and inhabited two different worlds. And I've always felt at home in either. The dawn of my life was dedicated to the process of physical exploration, where I actually moved about and *lived*. Eventually I managed to wander as far from home as I possibly could. But once I had returned from the moon and experienced that strange insight, I understood that this phase of my life was drawing to a natural close. Perhaps, similar to the American public in 1971, I had enough as well. Nothing I could do on earth could quite compare. I was home now. Anything else, any attempt at exploration for the sake of exploration, would quickly take on a deadening redundancy. After traveling so far, all other destinations seemed like pale shadows of a grander journey from which I'd already returned.

Somehow my attention was drawn down and in, deep into that vast realm of infinitely small spaces. The private experience of expansiveness that I had felt during our return from the moon in particular drifted into focus. There had to be some significance to it, something more than could be explained as mere elevated emotion, heightened awareness, or mountaintop experience, though I frequently used the latter term to describe it myself. The experience was too intense, too complete in its alteration of my sensibilities. It was somehow defining, but I was simply left puzzled in its aftermath. Something extraordinary had happened, and I didn't know what it was.

As the Congressional funds for the Apollo Program ran dry and the lunar missions drew to a premature close in the early 1970s, I turned my attention to the larger related questions about the basic nature of this "consciousness" we humans enjoy. The most neglected fields of consciousness studies lay in the realms of the mysterious states of mind that allow for epiphany and the psychic event. I read, groping in a way, for a satisfactory explanation. I knew there was something worthy of serious investigation if the issue was approached in a manner different from the traditional. I wanted to become intimate with the nature behind the phenomenon of human mental functioning—not neurophysiology necessarily, but rather the overall system that permitted us to evolve as we have. The question of how thought evolved seemed a more basic question than even how life itself evolved, though scientific tradition suggested otherwise. The reason for my brash certainty was simply founded: If the phenomenon of psychokinesis, as reported in the world's religious and mystical literature, had any validity at all, then the scientific doctrine called *epiphenomenalism* was a flawed concept.

Epiphenomenalism is the understanding that consciousness is merely a byproduct of physiological process; that it is secondary. Consciousness has no power to influence physiological processes, because it is the result of the evolution of our corporeal bodies. Epiphenomenalism is also mainstream scientific dogma. By this way of thinking, the mind cannot assume control over the machine, or the physical universe, though my experience on the way home from the moon seemed to suggest otherwise. There had to be some relationship between the intuitive experience of epiphany and the curious results obtained from the various phenomena being studied in parapsychology. Spread out before me was the vast internal landscape of a largely unexplored world.

The term *epiphany*, in the connotation of an intuitive insight, is certainly descriptive of what I experienced, as is the Greek word *metanoia*, which implies a change in thinking, even a new direction. But neither word individually, nor together, adequately describes the event itself. They seemed to surround the meaning without actually touching it. But there were other questions on my mind as well. I wanted to know what caused it to happen in the first place, what its place was in the vast scheme of things. Of course I wanted to know if others had similar experiences.

The latter question would take a decade to answer, and even then it would be answered rather indirectly. Engineers and test pilots have never been noted for introspection and spontaneous eloquent expression, so a direct approach would have been fruitless even had I known the right questions to ask. But it's significant that many of the men pioneering spaceflight began to express openly a more subtle side to their personalities after returning home. Several astronauts, notably Jim Irwin and Charlie Duke, immersed themselves in their religious callings; Alan Bean and Alexii Lenov, a cosmonaut, both found expression through art. Rusty Schweickhart spent a major portion of his subsequent career pursuing environmental concerns, often speaking eloquently on subjects closely related to my own interests. Al Worden published a book of poetry. I've found it interesting that all these Americans, with the exception of Worden, were lunar module pilots, which gave them a bit more latitude for contemplation on their way home.[1] And through the years I would learn that significance can be found in the routine as well. You don't have to journey to the moon to experience it. In the vague chaos of everyday life, ideas come to you in the middle of the night, in the shower, in dreams. Sometimes they are pulled together, and made whole, irrespective of the original sequence of their sources. They are life's little everyday epiphanies. Sometimes they can shape and alter a life forever.

During the weeks and months that followed the moonshot, I read literature on the nature of religious experiences, as well as the very limited scientific offering outside of religious and mystical writing that dealt with the nature of human consciousness. I also met with renowned psychics and highly intuitive men and women to discuss what it was that they experienced during moments similar to what I experienced. After a few weeks into this work, I knew I was on to something, though I still didn't know precisely what. At times I felt as though I was on the precipice of resolving a grand mystery.

Intuitive insights, ESP, and epiphanies, I knew, are just different names for perceiving information. As an engineer, I understood information as simply a pattern of energy. Consequently, it was evident that epiphany and metanoia are natural phenomena. I also came to realize that they are common to religious mystics and agnostics alike when considered in terms of information and how it is managed by an evolved organism (humans). It occurred to me that the ways of managing information in this manner could be similar to the way energy and matter are managed in nature; that is, increasingly complex forms of matter evolve increasingly complex forms of information. If the physical means driving evolution of information could be revealed, the significance couldn't be overstated: We would then comprehend what brought about such an ephemeral yet life-changing sense of understanding. Clearly this should have been, I felt, the domain of scientific inquiry, yet there didn't appear to be any serious effort being made at answering such questions. Evolution and even life itself has traditionally been considered to be driven by random processes and natural adaptation to the environment, not systematic processes. I was now convinced otherwise, but where was the evidence?

Mystics refer to such events as "religious experiences." Scientists scarcely address the subject at all, eschewing subjective events altogether. Somehow nothing I read seemed to capture the essence of what I wanted to know, and I realized that it would be necessary to form a new structure of thought for myself. At the same time, I wanted it to be consistent with the methods of science and still not ignore experiences reported by mystic traditions throughout the millennia. Yet I couldn't tacitly assume that either religious or scientific approaches to such events would necessarily corner the right answers.

A classic book of case studies of spontaneous expansive experiences written at the turn of the century by Dr. R.M. Bucke, entitled *Cosmic Consciousness*, set the tone for my own inquiries. Epiphany, I became certain, is a latent event in every individual. It is, to a large degree, what has allowed humankind to evolve in our thinking, as it brings about a sudden synthesis between existing ideas. Where mystics have believed the more startling insights to be a supernatural phenomenon, I was reasonably sure it was entirely natural, even normal; perhaps an emergent characteristic of ongoing evolution. Everyone experiences that potent, ethereal sense of *aha*, and for a brief moment, they glimpse the larger structure of a problem in their lives, resolve a conflict in their thinking, or glimpse the grand pattern of the universe itself. The idea of epiphany can be viewed as an abrupt organization, or reorganization, of information in a way that produces new insight at the level of conscious awareness. And that's what occurred, I believe, while on that fateful journey from the moon. I became quite sure of this. Yet I couldn't honestly describe it as a "religious" experience.

It is quite a different matter to suggest that an evolutionary product (the brain) spontaneously reorganizes its information to produce a new insight at the level of conscious awareness, than to assume that what one suddenly comprehends is the word of God. The latter, of course, is the more popular and common view of such events in a society steeped in traditional religious and cultural beliefs. All I wanted to know was why and how it happens. I wanted a more secular, scientific answer.

Bucke related in his case studies that these spontaneous events brought not only a more expansive viewpoint, a sense of inner peace and wellbeing, but also an unshakable feeling of immortality, accompanied by joy. This was a fairly precise, yet secular description of my own experience. It would have been quite easy, I suppose, to have fallen back upon some explanation such as having touched the face of God. But as a metaphor it didn't appeal to me, and it certainly wasn't literally true. I was convinced that these were natural events, not supernatural or magical, though certainly beautiful and profound. The ecstasy I experienced was somehow a natural response of my body to the overwhelming sense of unity I received. I saw how my very existence was irrevocably connected with the movement and formation of planets, stars, and galaxies—the ineluctable result of the explosion of an immensely hot and dense dot at the center of the universe billions of years ago. Or, if quasi-steady state theorists are correct, as it now appear they may be[2]—the ineluctable result of continuous matter creation in super clusters of galaxies.

Human volition, similar to human behavior, is rarely predictable, much less predetermined. I've had to make tens of thousands of choices in my own life, each of which would lead me down a different path. My own volition had given me a wife and family, and had taken me into the Navy, and finally to outer space. Human volition can also bring about the difficulties we deal with in life. In late 1971, it brought about separation from Louise and eventual divorce, as this new interest of mine moved toward an obsession. It brought the agony that accompanies the separation of loved ones when such difficult choices are made. The shape of an individual life seems drawn with a capricious pen as the consequences of our choices can never be fully known.

In the same unpredictable manner, the collective human volition would likely take humankind throughout the cosmos one day, unless we destroy ourselves first as a result of being unable to evolve beyond war-mongering. But to demonstrate, even to myself, that volition is real and not just a grand illusion, as scientific determinism decrees, this doctrine known as epiphenomenalism needed to be falsified. Here lay the crucial key to my bewilderment. It was this cornerstone of scientific thinking that distorted my own perception. As innocuous as the concept of epiphenomenalism may seem on the surface, it suggests the Newtonian idea that the fate of the universe is predetermined only by the laws of classical physics, and therefore entirely predictable. It allows no room for human intentionality. What this implies is that we humans are not really in control of our lives, but merely complying with the predetermined course of our fate as dictated by the immutable laws of physics. Surely, there must be an answer between a nihilist's materialism on the one hand, and paternalistic, supernatural deism on the other.

In the months following our return from the moon I began reading the mystical literature of both Eastern and Western religions. But I was careful in choosing my material, preferring esoteric literature, not exoteric (institutional) viewpoints. It was the "religious" experience, or epiphany, that occurs outside the influence of the Church and its dogma I was most interested in. Then one day an idea came to mind, one that would continue to grow for decades to come. From meager funds I commissioned a study by a qualified research team to dig up some facts on esoteric practices in various world cultures, and they came across some interesting discoveries that seemed to describe the essence of this epiphany. What the ancients, who wrote in the Sanskrit of India, described as a classic *savikalpa sainadhi* was essentially what I believe I experienced. In Eastern thinking, this phenomenon is a moment in which an individual still recognizes the separateness of all things, yet understands that the separateness is but an illusion. An essential unity is the benchmark reality, which is what the individual suddenly comes to experience and to comprehend.

I recalled so vividly the separateness of the stars and planetary bodies on the way home from the moon, but simultaneously knew I was an intimate part of the same process. This is the most salient recollection of the experience, and in a sense, defines quite precisely what I felt. Since the Newtonian era, classical science has described the separateness of physical objects. Demonstrating that they are inherently interconnected at the level of atomic structure was, in 1971, still a decade in the future. Somehow I sensed this flaw in Newtonian science, but at the time there was little experimental evidence to suggest it. The attribute of quantum interactions called *nonlocality* had been established in theory for 45 years, but had not been experimentally verified. Moreover, most physicists believed it played no part at the level of our macroscale universe, confined, so to speak, to the level of subatomic particles. However, I was confident that the wheel of science, however slowly, would eventually make its revolution.

In Idealist philosophies, where consciousness is the basic reality, perception of anything other than the basic unity of all things is called a *dualism*, a separation from the source of existence—the godhead. Separation causes pain, and reunion, by the same token, a heightened sense of joy. It is for this reason that all major religious traditions promise ecstasy when reunion with the Source of existence is achieved.

But modern psychiatrists and psychologists find a similar phenomenon. When individuals achieve a reconciliation with sublimated and disassociated memories, pain is released, healing takes place, and a greater sense of well-being and joy is experienced. In other words, the bridging or healing of sublimated memories (dualisms) results in the gaining of a greater perspective and a sense of wholeness. The reverse is also true: Gaining a greater perspective and a sense of wholeness, that is to say, learning, can be interpreted as bridging a dualism. This as well seems to bring with it a heightened sense of well-being. Certainly those humans whom we find most admirable in life are those who have achieved wisdom and possess an inner glow, a sense of well-being that bespeaks a fine-tuned integration of life's experiences.

An epiphany, therefore, if it results in a change in thinking, is a joyous experience. It is essentially the healing of a duality, which is what I experienced in the heavens. Or at least that's what I came to believe. Not only was there a sense of unity and wholeness with the cosmos, but also a duality, a schism between my early religious upbringing and my later scientific training, was suddenly and quietly reconciled to a meaningful degree. It was this new level of understanding that produced a potent sense of visceral *knowing*, the kind of understanding that you feel throughout your body as well as know with your mind. I also saw how the same process was at work in less dramatic ways in everyday life. Even the ordinary process of learning, as the brain synthesizes new information, is a related phenomenon. One need not go to the moon nor climb a mountain in order to experience epiphanies and metanoia. We all experience them in small everyday ways. When we establish the connection between our mood and what we had for dinner, even this is a small epiphany. It is nature's way of evolving our understanding.

As illuminating as this analysis was for me, it didn't fit within the materialistic philosophy of science, nor did it provide insight as to why or how the human organism came to function in this way. There had to be something more.

11

You bring your own history to the table when trying to make sense of such puzzling moments. The perception of experience is, after all, subjective. While putting myself through a rather rigorous course of reading, I contemplated my own formative years and religious upbringing. I saw how they could throw shadows over what I was trying to illuminate, how there was a history steeped in culture that colored my vision. My boyhood was a traditional Christian one full of wonderful ancient myths that largely dictated my sense of morality. And there were the cultural myths of the Old West, the pioneer myths in particular, handed down from great-grandfather, to grandfather, to my father, and finally to me. These were myths in the form of stories of self-reliance, stories of heroism on the frontier, of sin and redemption, the relationship between good and evil—almost all of which were overseen by a paternalistic godhead. They were stories that provided instruction, stories that subtly shaped judgment.

They were also stories that could lend a particular bias to my own interpretation of how the human mind and the universe itself are fundamentally structured. Though the myths that shaped and guided my formative years were useful to me as a young man, I saw how they fell short of explaining the questions I was now asking. Yet they were still the prism through which I tended to interpret the world around me. And they were a basis requiring accommodation. What I needed was a new myth, a new story that could more accurately describe a larger world, one surrounded by billions of other worlds, a story that was compatible with our 20th century, compatible with an emerging spacefaring civilization. I also realized that myth, when it is new, has always carried the label of Truth.

In the fall of 1972 I made a radical departure from what I had done for the past 20 years. I left NASA altogether and became a full-time student of that infinitely broad field of study that has been contemplated ever since the human mind was capable of self-reflection. I wanted to study the totality of consciousness. I strongly believed that consciousness, as a field of inquiry, encompasses all human activity. It fits precisely into the gulf between the way science looks at the world and the way various cultural traditions do. Mystical traditions assume, implicitly or explicitly, that consciousness is fundamental. Scientific tradition (epiphenomenalism) explicitly assumes it is secondary. It seemed to me that the study of consciousness provided the only unified approach to the questions of who we humans really are, how we got here, where are we going, and why. But I soon realized that the term *consciousness* itself has different meanings in different languages and cultural traditions.

I began infinitely hopeful that the methods of science would eventually provide the answers, and with only two fundamental assumptions. The first was that we were dealing with events likely explainable by natural processes, and second, that all human experience is valid, or real to the percipient. Only the interpretation or *meaning* given to the experience is subject to question. If supernatural or paranormal events were actually involved, that would emerge in due course. The vehicle for this project would be a nonprofit foundation that would allow me to function as an independent scholar. Academia wouldn't be very receptive to these interests, but I knew these were issues naturally requiring a multidisciplinary approach. Thus, the Institute of Noetic Sciences was conceived.

What I initially envisioned was an organization that wasn't so much a place as a state of mind. At such an institute scientists from across disciplines, and qualified lay people with similar interests, could come together to share questions, insight, and opinion, and then research and write on these subjects that were so close to their hearts. The forum would not be academic in the traditional sense. It would have an auxiliary purpose as well: funding key research that otherwise likely wouldn't attract money from mainstream sources. This, of course, would require tax-exempt status and reliable sponsorship. At the beginning, this didn't seem like an insurmountable task.

In the fall of 1972, an auspicious series of phone calls and introductions led to a meeting with a philanthropic couple living in California.

After these initial discussions they pledged an annual sum of $600,000, as they had decided this was critical work we wanted to accomplish. In the following weeks I contacted a small number of people I thought could really make this thing take off, and they generally responded with unbridled enthusiasm. We organized a lengthy retreat to finalize plans and to organize budgets and programs. The future appeared bright.

Then in January, after everything seemed to be sailing in the right direction and staff people were moving into place, I made a final visit to our would-be sponsors. When I arrived at their home I discovered their financial empire was in disarray; they told me their parent company had unexpectedly declared bankruptcy, and the funding they had promised was now out of the question. The institute had just opened its doors and its primary source of financial support was lost.

It was during these trying years that I picked up a new dictum: trust the process. Funding, I would learn, was always going to be a problem. But this was no reason to give up. With faith and perseverance the means would always come together so that we could take the next step. The institute's nascent years proved humble enough, as it was staffed primarily by myself, Anita Rettig, part-time employees, and devoted volunteers. Anita and I had begun dating after Louise and I had parted, and we both were devoted to the project.

This was a restless time, and in its own way, a romantic time. The days were fueled by a kind of idealism that propelled us into a vast unknown. To pay the bills we incurred and the small payroll, the institute was funded by whatever I could garner in lecture fees and the occasional donation of one or two thousand dollars by interested parties. Gradually a number of individuals were gathered to form the nucleus for a board of directors and a professional staff, and more projects were undertaken. All the while, Anita and I improvised. Anita had a gift for public relations, navigating those tricky waters of the media, and managed to find television and newspaper coverage for the institute, which eventually would translate into revenue. The donations the institute received in the beginning usually amounted to less than a quarter of what was needed each month, but in those early days, when it was essentially just Anita, myself, and a minimal staff, we seemed to be touched by an occasional miracle.

One rainy morning, a battered Volkswagen microbus pulled up before the plate glass windows of the office. A young lady in her early

20s came to the door and asked about the institute—what we were setting out to accomplish, how we were going to accomplish it, and why. We talked for a few hours, and as she got up to leave she asked if I would autograph a photo for her child. As I did so, she wrote out a check, handed it to me, and was gone as suddenly as she appeared. Once the microbus had pulled away, I glanced at the check and saw that it was written out for $25,000. Months later we learned that she was an heiress who wanted her money to go to worthy and humanitarian causes (this was the early 1970s, after all). Ours, she was sure, was such a cause. Such synchronicities allowed us to keep going.

But we didn't always have to rely on miracles. There were other resources to be drawn upon. In the fall, after opening our doors, I contacted Wernher von Braun, who was still working in the private sector, and asked if he would speak at a fund-raising dinner for the institute in San Francisco. There would be a host of other speakers, but he would be the keynote. He eagerly agreed, and through the course of the engagement we knew we had achieved both a public relations and financial success, though the get-together was criticized for having too much speech-making and not enough entertainment. But as a result of the evening we raised nearly $20,000. Once again, it looked as though our doors could remain open for a while longer.

Money is a curious thing. There is always the temptation to turn an organization into something that was never intended just to keep it alive. This is a corrosive phenomenon common to all underfunded ventures, and I was always afraid the institution would lose its idealism were we to indulge too enthusiastically in fundraising for the sake of fundraising. The Institute of Noetic Science was founded so that serious scientific work and discourse could be conducted concerning this elusive phenomenon we call consciousness, not merely to perpetuate itself. So we had to address this dilemma day in and day out if we were going to remain legitimate. We were successful, but in the beginning we paid dearly for our idealism.

The research we sponsored in these early years I am proud of to this very day. It was all cutting-edge, much of it considered "far out" at the time. But all of it would lay a solid groundwork for the future. Dr. Carl Simonton, an oncologist who was just finishing his military service, was interested in how individual attitudes and thought processes influence those afflicted with illness, particularly cancer. With the help of influential board members, we raised the funds to assist Simonton in his research. Brendan O'Regan, a biochemist working at Stanford Research Institute at the time, urged us to focus on health issues in general. He directed noetics efforts in the medical areas, and initiated the first research in several areas relating to health topics, such as the effectiveness of meditation and acupuncture. Today it is widely considered the earliest and best work of its kind. We were also enthusiastic about the biofeedback work of Drs. Elmer and Alyce Green. O'Regan became the institute's vice president of research in 1975, and worked in that capacity with unflagging dedication until his untimely death in 1992.[1]

As the decade rolled on, the relentless challenge of running the institute gradually took its toll. In 1974 Anita and I had married, and her children from a previous marriage were nearing college age. Karlyn was just graduating, and Elizabeth was still an undergraduate. As every parent knows, this takes money. In 1978 and 1979 I had to concentrate on meeting this need, so I tried to recruit particular individuals who could keep the dream alive, as it were, but with the intention of returning full-time as soon as possible. Along with Brendan O'Regan, I had met Dr. Willis Harman, who at the time also was working at the Stanford Research Institute. After a campaign of persuasion by the institute's directors, he finally came on board as president. Diane Brown (Temple) managed the administrative details for a few years until Winston Franklin, an executive formerly with the Kettering Foundation, was induced to join us. An auspicious combination of scientific interest, a proper understanding of fund-raising, and a solid administration were finally in place. So as the decade drew to a close, I was able to stay on as chairman of this diverse and devoted board of directors, while tending to my family's needs. Unbeknownst to me, however, a surprise lay in the offing.

With this capable and influential board, a membership program was undertaken, and interest in the institute began to spread. By late 1980, membership stood around 2,000. Publications were enhanced to better report developments, and more staff was added. But influential board members also tend to support projects corresponding to their own interests, and with a broadening of scope, I was concerned that focus would be lost. Funds could be raised to research the "channeling" phenomenon, "miracles," and the afterlife (if there was one), and a few of the board members thought the institute needed a more spiritual or mystical orientation. I was only mildly interested, and sometimes tended to be less than diplomatic in voicing my own opinion. These other subjects were interesting, but peripheral, and I believed they would all yield to a more direct approach through the sciences, so as to uncover the underlying electromagnetic and quantum mechanisms of brain and body. The mystical questions weren't the fundamental questions I'd been struggling with all these years. I felt, if we knew what really gave rise to our awareness and thoughts, and how they had evolved, the answers to such spiritual issues would be forthcoming.

Meanwhile, centrifugal force came to bear. Similar to all young organizations, there would be severe growing pains to be endured. In the end I paid for my stubbornness. In 1982 the board wanted a broader program closer to their own individual interests, and conspired to elect a new chairman. I suddenly learned that I'd been replaced. So I withdrew for a while to allow emotions to cool, in a way relieved of the administrative burden of managing a growing organization. But of course I was also hurt.

My life didn't exactly come to a grinding halt. If anything, it only grew more hectic as I spent more time in the private sector. At the same time, I wasn't about to give up on my studies. So in the mornings and evenings I continued my own private research—you do not necessarily need a laboratory to contemplate the nature of the universe. All you need is a room with a window, and perhaps a book or two.

12

It had dawned on me from the earliest days of the institute that there were two critical flaws in traditional thought structures that tended to distort the Western view of reality: evolution and intentionality. Or so it could be argued. Evolution as a general principle of the universe had already been demonstrated, but only in this century. Moreover, it wasn't yet accepted by fundamentalist and traditional thinkers. The remaining key lay in human intentionality. If intentionality in its strongest form of valid mind-over-matter demonstrations could be found to confirm mystical lore, and if it is a general human capability, then determinism and epiphenomenalism were clearly outdated ideas. This certainly seemed to be sound, simple reasoning, and so I set out on my own to discover which way the evidence would stack up—for the traditional way of looking at things, or for some new "myth" of our existence. I learned the lesson, however, that *simple* and *easy* are not necessarily synonymous.

There were fundamental obstacles to be surmounted. The really pivotal ideas concerning evolution and human intentionality had to be illuminated in new ways if epiphenomenalism was to be denied. Until the 20th century, conventional wisdom from all quarters accepted that the universe was basically in stasis, with change being only superficial whitecaps on an otherwise deep, still, unchanging ocean. It wasn't until Edwin Hubble discovered the expanding universe in the 1920s that scientists proposed, and seemingly verified in the 1960s, that a "big bang" of some sort initiated it all. In the aftermath of this discovery, evolution as a general rule for the universe was established. Darwinism was the door-opener, but it contained a critical flaw: Life had not likely evolved as a result of mindless, random mutation, but through an intelligent and intentional universal processes. The concept of a dynamic, continuously changing universe, of which our world is a part, fractured a major foundation block in most of the world's traditional thought structures. And now, just a few decades later, even though the big bang is under attack as being too simplistic a theory (see footnote 3 in Chapter 10), the concept of continuous random change and evolution is still solidly supported, but the time frame of galactic change and evolution is vastly increased.

The importance of human intentionality has been denied by science as well as most religious traditions. As far as scientific philosophy is concerned, it is basically an illusion, because reality is determined only by the movements of particles and energy in accordance with the laws of physics. If that were actually the case, then consciousness, of course, would be but a secondary product of the universal processes.

Cultural traditions derived from flawed interpretations of mystical experience disagree only about cause, not effect: The universe is governed by the supernatural choices of deities, and if humans want relief from suffering, they must supplicate the deity to affect change. When positive results are obtained, it is because the deities are pleased or compassionate, and "miracles" take place. In most tribal mythologies, capricious, angry, or loving, but certainly independent-minded gods, serve much the same purpose for which science uses the laws of physics and biology. The purpose of both is to provide some order in an otherwise enigmatic universe. In this way the world becomes more comprehensible. Unfortunately, the origin of the laws of science and the origin of deities are equally enigmatic.

When you look at any phenomenon and any accompanying paradox it contains, larger new truths arise from the collection of data. You see patterns, the trajectories of events. The same is true, I discovered, when you examine the history of human civilization.

It becomes all too clear that human mental processes, as measured by the capabilities and the knowledge base of our species, is growing at an incredible rate. As time has passed, our knowledge and capabilities have expanded exponentially. That is, they have doubled throughout decreasing periods of time. And this has been true for all of human history.

Take into consideration that in the millions of years since the first organisms sprang to life in the seas of our planet, only for perhaps the past 50,000 years have there been spoken languages that approach the complexity of modern languages. Anthropological evidence suggests the long road of language development may have begun between 1 and 2 million years ago with a change in the position of the larynx at the base of the human skull.[1] Certainly there were the grunts of pre-humans and primates, the howls of a dog, the whistles of porpoises and whales. And certain behaviors, such as courtship rituals, convey information to others as well, but by and large, precise oral language was nonexistent. Language has been but a recent and brief innovation in an otherwise silent world of signs and signals.

Of the last 50,000 years, there has been a written language for only the past 5,000. We humans have only been meaningfully self-analytical for the past 3,000. Just a few centuries ago Gutenberg made the first printing press, and not until the last century could man transmit a message through electrical means. The 20th century will be remembered for all time for three of the most remarkable human accomplishments: splitting the atom, space travel, and the advent of the computer. We are now simultaneously living in the Atomic Age, the Space Age, the Information Age, and the Computer Age. All four are, of course, unprecedented in human history.

But more importantly, the time period between doubling the number of new innovations and major developments has been reduced to just a few years, and doubling the amount of information available has been reduced to just a few months. Indeed, the periods for doubling the human population and the speed of transportation has been reduced to less than a human generation in my own lifetime.

What is most startling when considering this history is the extraordinary brevity of the history itself. The acceleration of change and the effect humans have had on the face of the planet provides common everyday evidence that human intentionality is a powerful force and that evolution is alive and well. It is also *increasingly under intentional human control*. This evidence begs the immense and inevitable questions: where was intentionality before the human period, and what lies in the offing for the next centuries? What will our civilization be like in 100 years, in 200 years, 500 years? Will humankind recognize the destructive effect it is having on our planet, and change its thinking and its actions toward a more sustainable future? As H.G. Wells once said, human history appears to be becoming more and more a race between education and catastrophe. Of course the clock will one day run out on these threats, either by social catastrophe, or through disciplined human action, though only the latter should be acceptable. Ultimately the promise of our technologies lies in our eventual escape from a dying planet.

Though the suggestion seems amazing even now, it would doubtless have appeared unfathomable to the Renaissance men of the 15th century that their progeny would one day surround the planet with a network of artificial satellites in order to photograph the mother planet, communicate at the speed of light, and spy on one another. The atom was still centuries away from being definitively revealed. What we take for granted in our daily lives would be considered some kind of legerdemain 500 years ago, punishable by burning at the stake. Had today's scientists been alive then, they would have been condemned as sorcerers and magicians by the Church and its laity—much as were Copernicus, Galileo, and Bruno. Likewise, what seems implausible now will certainly be routine in the coming centuries.

Perhaps the most crucial aspect of our own evolution is the degree to which we can be self-reflective. And this, too, has its own brief history. One of the most significant developments in recorded time occurred in approximately the sixth century BC, with the lives of three extraordinary human beings who lived almost contemporaneously, yet without knowledge of the others' existence. Each independently and brilliantly contemplated the mystery of being.

Gautama Buddha, the founder of Buddhism; Zoroaster, the founder of the Persian mystery school; and Lao Tse, who codified the Chinese wisdom of earlier periods and put them together to form the *Tao Te Ching,* apparently worked in relative isolation from one another, but produced the first recorded instances of critical, self-reflective, analytical thought about the nature of mind and being.

What I found so interesting in my reading, as many other scholars have as well, is that until then there hadn't been vigorous self-contemplation. The Neanderthal cave artifacts of ritual burials indicate the earliest documented concern with afterlife processes. The Neanderthal were a people who could not have engaged in meaningful introspective discourse 100,000 years ago, as their language wasn't sophisticated enough. But suddenly, beginning about 600 BC, there was this common phenomenon occurring worldwide. Four different cultures—the Chinese, the Indian, the Persian, and the Greek—all engaged in similar intellectual discourse about human nature and the role of mind, though reaching different conclusions and emphasizing different aspects of human mental attributes. It was as though human consciousness the world over had suddenly evolved to a new threshold of curiosity and understanding.

It wasn't until 399 BC that Socrates was offered the choice of exile from Athens or a death sentence for essentially asking too many unpleasant questions. In the end he chose the death sentence, to become a martyr for the right to free inquiry, which is exactly what he became. Socrates was widely believed to have brought philosophy down from the heavens, and the city elders feared he might inflict his questions upon the gods themselves. The Athenian government probably understood that once you start asking questions it is rather difficult to stop. But the genie was already out of the bottle. For the next 2,400 years, man would critically analyze his every aspect and his context within this strange world in which he found himself.

The development of language—spoken, written, and in the artificial languages of machines—parallels the growth of our technologies and our conscious awareness. The written word gave us an extended memory for use by our visual senses. Today we can read the ancient works of Homer, the ancient story of the Flood as it is recorded in the book of Genesis, and *The Epic of Gilgamesh*, and we can create revolutionary documents such as the Constitution of the United States to govern ourselves. More recently we have created extended memory for the auditory sense by inventing sound-recording devices, and later, multimedia devices that extend memory for replay to all our senses. But before this there were no man-made ways to record the human experience for direct use in the future. Yet storing information about experience is essential to any evolutionary system.

So we must ask the question: what are nature's ways? Nature provided the genetic code as one natural means to organize and preserve information for later use. I was beginning to suspect that there may be other mechanisms as well. But here we are getting ahead of our story, as those mechanisms were not identified for another two decades.

13

When I look back on the life I've lived so far, I see how it could be read as a small-scale metaphor for the course Western civilization has taken through the millennia. That is, my life and the progress of civilization have something of the same shape. In this century progress has been swift and severe, and my generation has seen the greater arc of its accelerating trajectory toward a global transition. We may be the last to witness this extraordinary human drama, as such acceleration clearly isn't sustainable, for a number of reasons, such as environmental tolerance.

When I take into consideration my austere origins on the plains of West Texas in an agrarian family whose sustenance was largely derived from manual labor aided by un-powered tools and domesticated animals, I am overwhelmed. To have begun in such a simple place and time, and to have seen so much, is at times incomprehensible to me. Likewise, when I consider the earliest agrarian humans on the planet and where we have come today as a civilization, I find the progress astounding. In 1930, space travel wasn't yet a significant part of the science fiction literary genre, but 40 years later I would travel in a spacecraft that would take two men to the surface of the moon while its sister ship orbited overhead.

In the largest sense, the evolution of humankind has made similar progress. We've evolved from a primitive species with limited knowledge with which to fashion tools, to a civilization capable of building machines that can split the atom or fill a magic box with image and sound. Yet our civilization is still in its chronological infancy; in geological time, just a few years out of the trees. As a species we seem yet juvenile, lacking in vision, unprepared for our own evolution, even blind as to the direction we are evolving. In this respect we lack any thoughtful, consensus judgment to guide our conscious volition, because we are still uncertain as to whether we actually possess volition at all, if one accepts the accounts of Western philosophy, theology, and science.

Late in 1972 I began taking such thoughts more seriously when a strange series of events occurred in rapid succession. Oddly enough, all of this happened just as I was preparing to leave NASA and about to open the doors to the institute. These were not events that I would consider mere serendipity, but rather events governed by the mysterious cadence of synchronicity.

In the fall of that year I traveled to Little Rock, Arkansas, to speak to a group at a convention—one of my first engagements of the sort. It promised to be a special occasion, as my mother was going to drive from her residence in Oklahoma to meet me. At the time she was having severe difficulty with her eyesight as a result of glaucoma, and without her glasses was legally blind. Through the years her glasses gradually grew thicker, as she considered corrective surgery too risky. And now she really could not see at all without them.

During the conference I met several remarkable men and women, one of whom was a man by the name of Norbu Chen. Norbu was an American who had studied the earliest form of Tibetan Buddhism, a form that was liberally infused with ancient Tibetan Shamanistic practice. He was a small man of quick movements, graying beyond his years, inscrutable, and always in the midst of controversy. He also purported to be a healer. One evening after an entire day of speechmaking, I introduced Norbu to my mother, who was at the time in her early 60s.

My interest was twofold. I wanted to find out whether Norbu Chen was real or just talk, and to help my mother if that was possible, though I was skeptical. My mother, being a fundamentalist Christian all her life, had definite and traditional ideas as to how the mind was capable of influencing matter through healing—either by the hand of God, or by that of Satan. There was no middle ground. Norbu did not think of himself as either, but was quite convinced he could help. Making no promises, he merely suggested that we try, and see what would happen. I was intensely curious, and my mother was at least a good sport about the whole thing. She agreed that something good might come of it.

The following day Norbu and I met my mother in the seclusion of my suite where he asked her to sit in a chair, remove her thick glasses, and relax. I watched from across the room as this strange Asian-trained man did what he claimed to have done for so many years. Then I witnessed my mother settle deeply into a relaxed state. After placing himself in a meditative trance (he claimed) through singing his strange mantra, his hands floated over my mother's head, pausing over the eyes. There seemed to be an unspoken acceptance on her part, a silent trust in this man she had never met until this weekend.

After a few minutes of this, Norbu gently announced that he was finished and suggested she go to bed, sleep well, and treat herself kindly, as though she had been through major surgery. His prescription for nourishment was grape juice and broth. As I sat there in the chair observing, there was the hope that I'd just witnessed the extraordinary. I wanted something to have happened, but at the same time I tried to be the detached, clinical observer, and not let my expectations soar. In any case, I didn't have to wait long for the results. At 6 o'clock the following morning my mother came rushing to my room, exclaiming, "Son, I can see, I can see!"

Without pausing to let me come to my senses, she proceeded to demonstrate her claim by reading from her thumb-worn Bible with glasses in hand. Then once again she said more quietly, "I can see. Praise the Lord, I can see!" Dropping her glasses to the floor, she ground the thick lenses into shards under the heel of her shoe. Needless to say, I was impressed.

I am not, by this account, nor with any other anecdotal story, attempting to convince the doubtful. That can only happen when the open-minded skeptic sets out for himself or herself to view (or better, to experience) such peculiar phenomena (at least peculiar to the Western mind), and conducts a careful investigation, unbiased by traditional interpretations. This wasn't science, but as far as I was concerned, it indicated where I personally needed to probe more thoroughly. All I can say is that it absolutely did happen in just this way.

Afterward I experienced the deep-down astonishment that arises from witnessing the extraordinary. This was an event I couldn't explain, but I couldn't deny it either. I knew my mother's reaction was authentic, and *she* hadn't been duped about her own sight. She proceeded to drive home alone, several hundred miles, without her glasses.

After this episode I was sufficiently impressed, so I invited Norbu to Houston for a visit so that I might learn a few things from him myself. He arrived a few weeks later to stay many months, during which time I came to know not only Norbu the healer, but Norbu the man. What I learned was notably unremarkable. He wasn't especially complex, just a fellow with a peculiar capacity to heal that he couldn't adequately explain.

A few days after returning home I learned another lesson that I wouldn't soon forget. After going about her routine for several days with nearly perfect vision, unassisted by contacts or eyeglasses, my mother called one day to ask whether or not Norbu was a Christian. His name was clearly derived from an Asian culture, which she suspected didn't likely coincide with her beloved faith. Though I didn't want to tell her, she was adamant. She absolutely wanted to know the faith of the man who'd allowed her to see again. Reluctantly, and perhaps ominously, I told her Norbu was in fact not a Christian, and the moment I did, the deep pain of regret was clear in her voice.

Her new sight was not the work of the Lord, she insisted, but that of the darker forces of this world. She was absolutely certain that Norbu, being of another faith, must be an instrument of evil. No matter what I said to her, no matter how I explained my own secular understanding of such phenomena, she would not be convinced. Her vastly improved eyesight was the work of Satan. Hours later, the gift slipped away and thick new glasses were required.[1]

I was both distressed and intrigued by this incident—distressed that such an incredible healing would be dismissed, and by my mother's agony in making this personal decision. But the intrigue, the fact that the sequence of events could occur at all, left an overriding impression. How could I have been so ignorant of something so important? It set me on the search for other persons similar to Norbu, and gave me clear indication that I needed to learn something more about the role and power of belief in our lives. Whatever the clinical implications, it was clear to me that one's internal life, the subjective life, had fundamental importance. This was something science didn't address; I had paid little attention myself.

But at the same time I recognized a need for caution. Though I subsequently encountered many healers with similar capabilities, I also encountered many frauds. I've learned through years of experience that health and well-being are a product of total lifestyle. There is no panacea for illness in healers, allopathic medicine, naturopathic medicine, chiropractic, nutrition, and the like, though all can help.[2]

Looking back on these times, I see how naïve I was. For several years I would continue to underestimate the power of belief in our lives because of the pervasiveness of my classical scientific training. It still puzzled me that belief could affect anything at all. But I suppose naïveté was also in large measure the impetus behind my founding an institution where research I thought important could be carried out. I believed that if other scientists witnessed such legitimate phenomena in controlled environments, they would see that it was at least worthy of further study and become excited by the prospects. But there were invisible veils that such unbridled idealism could not see. As it turned out, disbelief was one of them.

It was my opinion then and it is my opinion today that disbelief prevents one from seeing what one wishes not to. My belief in the rationality of science blinded me to the equally rational consequences of disbelief. At the time I still suspected there might be a nonphysical component to consciousness, capabilities that cannot be attributed to physical laws. But more likely there were physical principles yet to be discovered. Whatever the answers, they would surely be revealed one day by a rational, thorough approach to the issues. These were natural, not supernatural events, well within the domain of scientific inquiry, and when validated, the impact on science would be revolutionary. But it should also change the way we addressed religion, philosophy, government, the way we saw ourselves in the universe, and the values we adhered to in daily life. Unfortunately, there were precious few who took the field of study seriously, as a number of eminent men of science had blunted their swords on these issues during the past century. But again synchronicity would arise in my own exploration.

In that same year, 1972, I was introduced to another psychic, a 25-year-old Israeli by the name of Uri Geller. At the time, Uri was unknown, lived in Israel, and had been brought to the attention of Dr. Andrija Puharich, an American physician with a quixotic turn of mind and several medical inventions to his credit. Similar to myself, he was looking for answers to puzzling human phenomena. After months of observing Uri in Israel and witnessing his extraordinary ability in realms of telepathy and psychokinesis, Puharich called to see if I would be interested in arranging some controlled studies in an American laboratory.[3] I told him I was, provided I could personally do some preliminary studies to assure myself it was worthwhile. Here again was my own skepticism asserting itself.

Puharich claimed that he had witnessed some positively amazing demonstrations of telepathy, clairvoyance, and psychokinesis—in fact, a very broad range of events labeled as parapsychological phenomena. He insisted they were the most convincing he had ever seen, and that Uri's incredible abilities demanded our attention. In later years Uri would work with many scientists and then go on to make a modest fortune working for oil, gas, and mineral companies, successfully dowsing for deposits that lay deep within the earth. The fact that such companies don't pay good money for failure is strong testimony to Uri's psychic prowess at least as a dowser. But in 1972 he was just a young, impecunious Israeli, finishing his obligatory service in the army and claiming he had been doing these feats since childhood, not initially recognizing they were bizarre—a characteristic I later encountered frequently. After our first meeting, which lasted several days, I was adequately convinced (in fact overwhelmed) that his abilities were real and not simply showmanship or a magician's tricks, that I promptly began to arrange for financial sponsorship through Judith Skutch and Henry Rolfs, who would later serve on the board of directors of the institute, and to arrange for laboratory tests to be conducted. Uri agreed and came to the Stanford Research Institute near San Francisco in the late fall to participate in several experiments that would demonstrate his parapsychological abilities under our auspices. A team of scientists would direct the work and observe the results. This, I thought, would help prompt real interest among established scientists, provided Uri could deliver on his claims. But it wasn't Uri who would disappoint us.

The key to good science in this field is always to keep the experiment totally under the control of the investigators and to use blind and double-blind testing procedures wherever possible. Because one is dealing basically with subjective events, it's generally more efficient to let the subject demonstrate what he or she does best, then design rigorous protocols that take advantage of the ability. It was in this spirit of cooperation and open inquiry that Uri and I got down to business. He even agreed to restrain his usual showmanship and flamboyance.

The resident scientists hosting the experiments were Dr. Harold Puthoff and Russell Targ, both widely respected physicists at the

Stanford Research Institute. Both men had made their reputations in more conventional fields of science, but they too were convinced that this was something that demanded their attention. They designed many of the experiments we would conduct to test for Uri's abilities. On several occasions they were aided by Dr. Wilbur Franklin of Kent State University, a metallurgist who examined many of the materials used in the experiments. They had each been working in parapsychological studies for some time now, and were well experienced in this type of research. Targ was also a skilled amateur magician, which blunted the first claim of critics—that "psychic stuff" was just smoke and mirrors. The first effort of the researcher is always to determine validity of the event. But the important work, once validity is established, begins with the search for clues to the mechanisms permitting these phenomena to occur so that a workable theory can be constructed.

One type of experiment Puthoff and Targ created is called *remote viewing*, in which random "targets" are picked, and the subject (Uri, in this case) is to describe the unknown object.[4] Typically, these targets would be chosen by invited outside scientists, sealed in an envelope, then selected at random. Uri, in a room all by himself, where he was isolated from receiving any possible information, would attempt to describe the setting. We found he could do just that.

Nearly every time Uri was given a target, he would promptly draw—quite recognizably—what lay there. This type of experiment was usually conducted in a double-blind fashion so that no one knew the correct answer before the tests were complete and checked by impartial observers. In some variations of this experiment a Faraday cage was used, which isolates normal electromagnetic signals. Puthoff and Targ went on to conduct the same experiment with dozens of other people, both those claiming psychic abilities and those not. Eventually they discovered that most any willing person with a bit of training could get significant results, supporting the idea that this was not only a natural function, but a common one as well.

It was also discovered in subsequent tests that the brain waves of two individuals separated and isolated by a Faraday cage could synchronize their brain waves. A light pulsed in the eyes of one would cause a certain EEG pattern. The second person, by merely thinking of the first person, would suddenly acquire the same EEG pattern. Somehow there seemed to be some sort of communication occurring between the two that we didn't know was possible.[5]

Another version of the same experiment was a telepathy test conducted with an EEG connected, which proved statistically significant. The brain waves of the percipients showed a marked change a few hundred milliseconds before the percipients reported an answer. Conscious awareness hadn't received information until nearly a half-second after subconscious processes had received the signal. The result is quite similar to tests of the five normal senses, in which conscious awareness lags behind subconscious signal processing.

In the initial plan we wanted to test all of Uri's special abilities, particularly those that demonstrated a strong component of psychokinesis, because that was the most bizarre and difficult to explain within the existing framework of science. One of his trademark capacities was the ability to bend common metal objects, such as forks and spoons. Of course one of the objectives was to test this while filming with videotape and 16-millimeter film. But when we placed a spoon on a table under a glass jar, we found he couldn't bend the utensil. However, when we allowed him to touch the spoon, he explained to us how it seemed to "turn to plastic" in his hand. But this was generally unconvincing to scientists who specialized in other realms of study. They claimed Uri merely had extraordinarily strong fingers and possessed the ability to twist the metal under his peculiar grip, or that he had some unknown solvent on his finger that softened the metal. Yet no one was aware of any such solvent that could be used in this way; the physicians in the group couldn't explain how he could be capable of twisting the metal so adroitly into such a neat little coil by merely touching it with a single finger. But the objections persisted. At times their explanations were more far-fetched than the event itself, and seemed little more than tortured rationalizations.

Most convincing to me, however, were the dozens of children I investigated who had watched Uri bend spoons in this manner on television. Shortly following the tests at SRI, Uri made a series of television appearances, during which he did his thing before the cameras. Soon after the broadcasts my phone would ring with frantic parents reporting that their children were bending spoons as well. I could usually sense what part of the world Uri was in by the parents calling to report that their children were bending the family silverware.

I went to a number of homes around the country, sometimes with my own spoons in pocket, or I would select one at random from the family kitchen. Typically it was a boy less than 10 years of age who would lightly stroke the metal object at the narrow point of the handle while I held it between thumb and forefinger at the end of the handle. The spoon would soon slowly bend, creating two 360-degree twists in the handle, perfectly emulating what Uri demonstrated on television. No tricks, no magic potions, just innocent children (with normal children's fingers) who had not yet learned that it could not be done.[6] The evidence continued to mount in this way, suggesting that these strange capabilities were quite natural and likely common in humans, though latent and seldom manifest. It occurred to me that we were possibly seeing the emergence of an evolutionary attribute, or the residue from an earlier one that was now fading. The most modern evidence (discussed in following chapters) suggests that these are latent capabilities in all humans, rooted in evolutionary progression and waiting to be developed and exercised like any other attribute or skill, such as artistic, musical, or athletic abilities.

During the six weeks we conducted formal experiments with Uri, there were also an incredible number of equipment failures and downright strange occurrences that no one could readily explain. Video equipment that he had no access to would suddenly lose a pulley, which would later be found in an adjoining room. Jewelry would suddenly be missing, only to be found locked in a safe with a combination Uri could not have known. There were literally dozens of such events. No one could explain what was going on, though we all had our pet theories. The occurrences were reminiscent of "poltergeist" effects reported throughout the ages, which many modern investigators have come to associate with emotionally distressed individuals, oftentimes adolescents. In spite of the equipment failures, however, we were successful in quite a number of our experiments at SRI.

In one we placed a large ball bearing under a glass bell jar on a table. We wanted to see if Uri was capable of moving the ball bearing without touching either it or the table. And of course we wanted to record the experiment on videotape. When we explained to Uri what we had in mind, he nonchalantly approached the table and placed a hand over the jar, swearing from time to time as the shiny gray orb sat motionless on the table. As the clock on the wall ticked away, and the repaired video equipment hummed in the background, the ball bearing refused to budge, until finally, after closing his eyes and raising his face skyward, it began to jiggle, then roll this way and that. As his concentration seemingly waned it slowed, then stopped. Finally, we had something legitimate on tape.

When we went to review what we had all seen in person, we were relieved to see that the event was caught cleanly within the camera's field of view. Nothing had gone amiss. However, the feat was still greeted with skepticism when our colleagues in science viewed what everyone who had witnessed the event and had thought monumental. They became red in the face, and some left, refusing to ever return to the lab. They accused Uri of being a fraud, and the rest of us of being chumps in an elaborate charade. But their accusations flew in the face of the solid scientific work that had been done, and I believe they knew it. Even some visiting scientists who watched positive results directly and in person angrily rejected what they saw.

With some experience in this business it actually becomes quite easy to detect the charlatans and frauds, and there are quite a few. Uri, however, was not one of them, though many times he could not produce positive results within the rigorous controls that we imposed, simply because we forced him out of his comfort zone with our requirements— all sensitives report that they need to achieve a particular internal feeling state to be effective at exercising their talent. A distracting environment, hostile people, or any number of other influences can make the necessary state difficult to achieve. Artists, musicians, great thinkers, and psychics seem no different in this regard.

14

By 1973 I was beginning to understand why normally rational people reacted as they did when they witnessed some of the experiments and viewed the evidence we had carefully collected. I believed then, as I do today, that it comes down to an individual's belief system, which each of us constructs through the course of our lifetime. This psychological construct is really just stored information we gather through the years and weave into a "story" that has both conscious and subconscious components. We use it to represent our reality. At any moment, the sum of the information and the meaning we attach to it is derived from all of the experiences previously accumulated; sometimes seemingly insignificant experiences. Our beliefs cause us to see the world in a way that is unique to each one of us, as it is quite impossible to duplicate each step in a life's journey. This is the source of our "story."

We often sublimate information and its significance in the subconscious, though it continues to influence our thinking and behaviors. Whatever our beliefs, we are likely to label them as Truth, and consider them permanent and absolute, because our internal representation is the only map of reality we can know. When presented with new information that challenges what we currently believe, reflex prompts us to attack the new in order to maintain the integrity of the old. We don't like to be wrong. Fears are aroused and the primitive fight-or-flight response typically summoned. Whatever our beliefs, they are comfortably familiar to us, like old shoes. I saw how the phenomena Uri Geller demonstrated could be seen by others in the scientific community as a reality at some level of awareness, yet immediately rejected as a threat at another. Any type of psychic event under conditions that gave it validity appeared to shatter accepted scientific thought—similar to a scientist earlier in this century who criticized J.B. Rhine's work with the statement that he wouldn't believe it even if it were true.

In the Newtonian worldview derived from Descartian philosophy, belief about physical reality is irrelevant, because belief represents only lack of knowledge. The physical universe was conceived by Newton as a deterministic machine grinding inexorably toward whatever destination God had in mind. The universe was simply there. It was absolute, and man was only the passive observer who needed to discover its physical laws in order to understand it completely, and thereby better follow God's commandments. Enlightenment, in this system, simply means belief in God and complete knowledge of how His creation, the physical universe, works.

By the mid-19th century a strict materialist philosophy, energized by Darwin's theory of evolution, had captivated most European and American intellectuals. The spiritual portion of Descartes' dualism was discarded. Indeed, philosophy itself nearly vanished from universities at the beginning of the 20th century, as scientists came to believe that all the problems of the physical universe were nearing solution. Then Einstein suddenly proposed that light was both wave and particle. This opened the way for a new physics—quantum physics—which has occupied physicists for the entire 20th century and totally revised our modern scientific thinking. But Einstein too was greeted with skepticism and censure, in the beginning.

Before the last century, life moved at a pace measured in lifetimes and centuries. In each culture the fundamental beliefs about the world were instilled in children before puberty, and mostly by family. In a largely illiterate and slowly changing world, the beliefs learned at mother's knee served for a lifetime, as there was seldom any need for young people to update the way they interpreted the world around them. There was no desire to change because cultural beliefs were considered absolute knowledge: Formal education merely added detail and reinforced the basis for the cultural beliefs. But in today's rapidly changing world, this process isn't enough. Beliefs of the past, though considered absolute Truth in their time, we dismiss as ancient myth and medieval superstition. We tend to forget that our forbearers were sincere and intelligent folk trying to understand the mysterious world around them. All they lacked were the tools of our time. We also tend to forget that we don't have all the answers today, as the search is incomplete.

Today we know that one's belief system begins its development prenatally, and unconsciously. Sounds, sensations, and feelings are stored while still in utero. When an infant struggles to find nourishment, and finally does so as it discovers its mother's nipple, even then the child is developing its belief system. It creates meaning for its mother, as it has found her soft, warm, and nourishing—a gestalt of information formed from many small acts, movements, and sensations. As the child grows it never loses this information about its experiences with mother. The memories are forever sealed, though some in abbreviated form or outline, mostly in the subconscious long after she is capable of fulfilling its needs. And so our beliefs accumulate moment by moment and experience by experience throughout our lifetime. Modern studies indicate that some memories are retained in exquisite detail, while others only sketchily, and some not retained at all. The emotional effect of an experience and the meaning attached to it at the time are factors in the amount of detail submerged in our long-term memories.

This leads to the point that our beliefs are our map of reality. We do not perceive reality directly, but only the information our senses present to the brain at any given moment, which is then compared with the existing remembered experiences to obtain meaning. Because this map is the only reality we humans know, we often make the mistake of thinking that our map is reality itself, when in fact it is just an incomplete portrait painted from memory.

This creates a problem, as all protocols and investigative procedures in science are designed for external measurement and validation. There are no protocols and procedures within the hard sciences for investigation of subjective events. Indeed, science eschews subjective data altogether. The technical term for such protocols is *epistemology*, yet there isn't a proper term for studying subjectively generated phenomena, particularly those that challenge long-established physical theory.

Dr. Willis Harman, who eventually became president of the Institute of Noetic Sciences, accurately put his finger on this aspect of the credibility problem for modern research on psychic events. Science has traditionally dealt with "objective" reality, accepting the Newtonian belief that matter could be studied independent of mind. In the 10 years before his death in 1997, Harman set out to change that by proposing what a proper epistemology need include. This is of particular importance because all observations are essentially subjective events. There are no truly objectively observable events in the purest sense of the term. All observations are ultimately subjective information organized in our brains, to which we attach a meaning as a result of experience and our accumulated belief system. Repeatable observations and measurements are deemed "objective," though they actually represent but a consensus of the observers who share a common belief system. And, of course, the consensus may be in error. If science is ever to have a complete understanding of how the universe is structured, it must include why and how it is that we "know." And, of course, the subjective experience is at the crux of the issue.

Because of this epistemology problem and the overwhelming disbelief of reviewing scientists, much of the more startling work at SRI was never published. The politics in the peer review system of contemporary science inhibits work that is very far from mainstream, and all but assures that such work will be ignored.[1] There were, in fact, some individuals who subsequently made careers out of debunking Uri, though they were, in my opinion, reinforcing their own self-deception, and perpetuating a flawed consensus. The argument that finding a possible way to simulate an event is proof that it could not have occurred as reported, is specious. But the heat to which the individual scientists behind our work were subjected was fierce, and even SRI was criticized as an institution for its involvement. Yet a considerable amount of psychic work was subsequently done there covertly under classified programs for various government intelligence agencies. This was in some measure instigated because the Soviet Union had quite an active and effective psychic research program in the 1960s and 1970s directed toward intelligence activities. These were, after all, the Brezhnev years of the Cold War.[2]

Among the very strange things that happened in this period involving Uri Geller and Norbu Chen were some personal events that made the greatest contribution to my own understanding, although they fell short of being science.

Norbu Chen was still in Houston when I returned from California. When he asked me where I'd been, I told him I was with Uri at SRI. He seemed a little put-off. Norbu, like all of us, had an ego that did not like to be outdone, so he told me that if I really wanted to see psychokinesis I should take off the heavy gold ring I was wearing and hold it loosely in my hand. When I did so, Norbu appeared to concentrate and passed his hand above my closed fist several times. Then he asked me to look at my ring.

What had been a fine piece of gold jewelry with my birthstone as a setting 10 seconds earlier was now bent, twisted, and impossible to wear. He hadn't touched it, and I had only held it loosely, but now it appeared as though it had been crushed in a vise. Norbu Chen was clearly a very powerful man.

I would have taken Norbu on my next trip to SRI, but funds were short for one set of experiments, not to mention a second, and the tightness of the controls necessary to get publishable results were downright oppressive. This frustrated Uri just as it would have Norbu. Moreover, whenever we got good, solid results, scientists would not believe them anyway, so taking Norbu to California seemed like a bad idea in any event.

The next time I returned to SRI with Uri, however, I was better prepared. Somehow I could anticipate the moment something important was about to happen. I also learned how to set a less daunting stage so that Uri could feel more at ease. Because he purported to be able to perform not only psychokinesis but telekinesis (the ability to transport a material object by mental means) as well, I challenged him one day to recover a camera that we had jettisoned while on the moon in February nearly two years before. It had a serial number recorded somewhere in NASA files, which I didn't know. Of course, were that camera suddenly to appear, it would have been a valid telekinetic event. At least *I* would know it was valid. Somehow this seemed to spur a flurry of strange happenings.

As we were sitting at a table in the SRI cafeteria a few days after this challenge, Uri asked for a dish of ice cream for dessert, which the waitress brought to him a few minutes later. After the second or third bite, Uri cried out in pain, and then blood seeped from his lips. From his mouth he took a blob of ice cream with a tiny metal edge protruding from it. He handed it to me, and I washed it in my water glass in full view of the seven or eight of us at the table. What I discovered was a silver miniature hunting arrow mounted over the silver image of a long-horn sheep, the sort of emblem an archery aficionado might have for a tie clasp or a medallion. I was utterly taken aback, utterly unprepared for what I recognized.

Though I've never been an archer, I had been given a tie bar with such an emblem a couple of years earlier when I visited an archery vendor's booth at a trade show. But I lost it shortly thereafter, along with an entire box of tie pins and cuff links during one of my frequent trips to and from Cape Kennedy in support of *Apollo 16* the year before—long before I'd ever heard of Uri Geller.

We all nervously chuckled at what had just happened, then returned to the lab for an afternoon's work. But the oddities continued. While momentarily alone in the little lab, I heard the strike of metal against the tile floor outside. I turned just in time to see Dr. Puthoff pick up something small and shiny. He didn't know what it was or where it came from; it just seemed to fall out of nowhere and land at his feet. When he handed it to me I saw that it was the tie bar that matched the emblem that had appeared in the ice cream. Even the broken solder joint matched, though when I had last seen the two, they were one piece. The atmosphere was growing downright eerie.

After some nervous laughter, Puthoff and I walked into the lab and began working with the apparatus for another afternoon's experiment. As we stood by ourselves at the laboratory table we both caught a glimpse of something as it dropped between us to the floor. After a moment of bewilderment, I reached down to pick it up. Here was a pearl tie pin my brother had given me as a gift after his military duty in Okinawa, which I'd kept in the same lost jewelry box. Three of Edgar Mitchell's lost articles recovered telekinetically within the span of 30 minutes. But no camera. Startling phenomena, but accepted science still remained just outside the door. (I still have the tie devices and the ring Norbu mangled as part of my cherished keepsakes from the period.)

Such bizarre events were typical of how things occurred that fall at SRI. They were often deeply significant to a particular individual, because they generally involved something personal, as with my ties pins. Though astounding to those present, these were not replicable events (and thus not good science). It's important to point out that the last two events involving my tie pins did not seem to come from anywhere, but rather just *appeared*. I groped for something to which to compare the phenomenon, and finally realized it was not unlike the behavior inherent to the electron that jumps from one level of orbit to another around the nucleus without traversing the space between the two. This was the only analogy I could come up with. In my own scientific studies, I knew how this created great consternation among the framers of quantum physics in the 1920s. So a question arose: Could it be possible that we were seeing quantum jumping on a macroscale?[3]

In the spring of 1973, I met Uri in New York to accompany him for an appearance on the Jack Parr television talk show. During the broadcast we were both surprised when Parr's stage assistants unexpectedly handed Parr several large steel nails. Parr enclosed them in his fist with only the tips barely protruding, and asked Uri if he could bend them. Without hesitation, and in much the same way that Norbu had passed his hand over mine as it held my ring, Uri placed his over Parr's, and concentrated. A moment later he asked Parr to take a look, and when he opened his hand Jack Parr's face turned ashen. The tip of one of the nails was bent about 20 degrees, whereas all had been perfectly straight just moments before. An awkward silence fell over the set; there was hardly anything left during the remaining minutes to talk about. At least this had been caught cleanly within the crosshairs of a camera.[4]

During my final days in Houston I invited a number of physicians to observe Norbu as he worked his "magic." Dr. Ed Maxey from Florida and a number of NASA flight surgeons helped with checking records and observing events. One event in particular was especially poignant because it involved Anita Rettig, and stimulated her interest in helping me organize the Institute of Noetic Sciences. A few months later we would be married for 10 wonderfully hectic years, but at the time she was ill with a kidney disease that threatened to require a lengthy series of dialysis treatments. As a last-ditch effort before admitting herself into the hospital, she agreed to have Dr. Maxey fly her to Houston for a session with Norbu.

She was frightened, pale, and very uncomfortable when she arrived, but gamely went on Norbu's diet regimen of grape juice while the doctors assembled, checked her records, and watched Norbu initiate his treatment. He knew nothing of her disease, and had no access to her medical records, yet after a few moments gave a correct diagnosis, which the physicians confirmed. He then proceeded to relate to her privately how and when the problem first came about. Only Anita knew the history that he related, but she later confirmed that the events had in fact taken place as he suggested. After about 20 minutes in his meditative trance, Norbu ordered her to sleep for the night. Tomorrow, he said, she could fly home, but she was to treat herself kindly for a few day and drink more grape juice. Her problem appeared to clear up immediately, and dialysis was never required. Her local physician, who had recommended dialysis, when finally informed of the procedure with Norbu, exclaimed, "Isn't it amazing what grape juice can do?!" Almost two years later she had a thorough check-up by an eminent internist in San Francisco, who found no trace of the kidney disease.

Uri Geller and Norbu Chen were the most accomplished psychics I have ever met, but only the first to demonstrate for me such powerful effects. In my mind, the evidence of their strength lay in the wide range of strange capabilities they could display upon request with quite consistent results in spite of constraints, sometimes unreasonable ones, which we imposed on them in the name of science. By being in close quarters with both Norbu and Uri for weeks at a time I gained insight into how they functioned as human beings.

What I discovered was that they were neither Satanic nor divine; just two regular men with impressive talents that science claimed they could not possess. Each was pleased with his prowess, and frustrated at the doubts and public controversy. Yet they helped me reveal what I needed to know, and provided a rigorous benchmark against which to measure these capabilities in humans.

For many years after these initial experiences with Norbu and Uri, I continued to find, or to be found by, hundreds of people around the globe who could utilize their strange abilities in powerful ways. And their explanations for the source of their talent generally conformed to their cultural beliefs. Some were in religious orders, some in remote locations, others in more "primitive" cultures. Some were ordinary Western folk previously afraid to talk openly of their experiences for fear of ridicule—or more importantly, fear of losing their livelihoods. We've just stopped burning witches, but we haven't stopped punishing them.

I soon lost all interest in accumulating additional data of this sort, preferring to work in splendid isolation at piecing together some structure for how the mind could produce these effects, and still be compatible with the detailed picture of physical reality that science was continuing to unfold. At the same time, I couldn't ignore the insights of centuries of holy men who had explored the realms of consciousness without the benefit of the marvels of modern science.

As I pondered these events, the puzzling questions that arose on the way home from the moon reasserted themselves. I recalled how in space certain truths seemed so brilliantly clear, truths that somehow became obscured within the atmosphere of Earth. In space it seemed so obvious that the processes of the universe are innately and harmoniously connected. But with the hubbub of daily life, this wasn't so clear at sea level. Our understanding and treatment of the psychic event, which exemplified for me a key to the bridge between mind and matter, had never been brought into the larger picture, being deemed either supernatural or phony. I knew the events were real, and yet our two major thought structures, science and religion, were mysteriously incapable of addressing the subject adequately. The flawed conclusions in our cultural beliefs were derived from flawed assumptions and interpretations in both science and theology, which sprouted from the dim recesses of history when the knowledge base was so frail and thin. Both religion and science were understandings evolved by our ancestors as they struggled to form a picture of the mysterious world around them, a picture we are still evolving today, making clearer, broader, more complete. However, we are still working at this noble task, though using many of the same flawed assumptions. This notion only reinforced my decision not to accept at face value any cultural dogma, but to reexamine all experience with fresh eyes.

It dawned on me that we were dealing with only two basic categories of events, rather than the long list proposed and used by early parapsychologists (such as ESP, telepathy, and clairvoyance). Even today I cringe when I hear those terms used to indicate some strange or abnormal capability. The two major categories relate to active and passive uses of energy, and are associated with our everyday functions of awareness and intentionality. When an individual is quiet, relaxed, and receptive, he or she can become more sensitive to and aware of energy in the environment, and of its patterns that we call "information." Naturally gifted and/or well-trained people are more aware of energy patterns from both internal and external sources than others. So the categories of telepathy, clairvoyance, and the like just denote a greater awareness of naturally occurring and man-made patterns of energy.

Intentionality is the active process of desiring or intending an action. Action requires the movement, or transformation, of energy—something each of us does every moment of our lives, even in such common events as lifting a finger. Psychoactive people, either naturally or through training, have a greater range of actions they can intentionally and directly initiate with their mind. I was coming around to see that there is actually nothing more complicated conceptually about these unusual processes than that. Psychoactivity is merely a means of managing energy and the patterns of energy that we term *information*. However, there are many subtleties involved in training oneself to better employ the process, particularly outside the body, and considerable complication in explaining how it fits within the theories of physics. There is also still considerable mystery as to how the brain actually accomplishes these feats. The processes are not simply the mechanical processes of gears and levers, nor the electrical processes of moving charges, as classical theories would have it.

I suspected early on that when these capabilities are utilized, they require no moral or ethical considerations different from those of ordinary awareness and intentionality in daily life. That is, the morality and ethics required are just those contained in the belief system of the practitioner. Psychoactive people can be wonderful and saintly or scurrilous and demonic, just like the rest of humanity. The energy involved pertains only to its quantum mechanical properties. Similar to electricity, it can toast your bread or power an electric chair. It requires no special dispensation from supernatural authority. And it is precisely for this reason that virtually all the world's established esoteric traditions require practice of self-discipline and acquisition of "spiritual" values first, allowing the psychic capabilities to manifest when and if they naturally emerge. The idea here is to have a more compassionate and wiser individual in possession of such abilities. The more ancient rites, such as the Voodoo of Africa and Haiti, and the Shamanistic practices of South American tribes, routinely use the capabilities against enemies as well as for friends. The morality is that of personal and tribal survival.

Years ago, Puharich carefully investigated a case of an alleged Shamanistic vendetta against a South American tribal woman. She was an outcast, banished because of a family dispute, and had apparently fled her village so that she might escape the Shaman's spell. Apparently her flight was to no avail, as shortly thereafter she was afflicted with dozens of tiny needlelike metal shards deep within the flesh of her body. They required surgical removal and left a welter of scars, but continued to reappear for a period, seemingly spontaneously.

After having seen what I'd seen and having heard what I had heard from credible scientists, I grew certain that these increased levels of psychoactivity are most likely latent, evolutionary, and emergent in our species.

But if our belief system will not accommodate these natural abilities and they are suppressed early, they will not naturally emerge in the individual; there is just too much dogma in the way. The physicist Max Planck nearly a century ago said that new ideas do not prevail by convincing the skeptic, but rather funeral by funeral.

Invisible Realities

Following these experiences with my mother, Norbu, and Uri, it was clear to me that not only individual belief but also human intention plays a large, though quiet role in shaping our lives. Obviously my mother intended to be healed, but not by what she feared were tainted forces. Uri struggled to bring back my camera but couldn't quite pull it off. Norbu intended to heal others, and, unfortunately, deformed my ring in demonstrating his range of capabilities.

Regardless of how one twists and turns deterministic theories of nature, such theories cannot accommodate these events, as both volition and intentionality are denied in the old scientific framework. My experiences with children performing psychokinesis didn't help validate the supernatural theologies either, unless God was asleep at the switch and allowing Lucifer to run the show. That meant the answers lay in natural realms, likely within quantum mechanics and some new theory of mind.

The story of quantum mechanics in the last 100 years is of particular importance to notions about consciousness. On the surface it may seem strange that the nature of subatomic particles and light could be entangled with functions of mind. But when we eventually understand the full story of consciousness, it will have to be compatible with our knowledge of how the physical universe is structured. And quantum mechanics is a key map to a strange and invisible world that underpins everything we know. Research carried out in the last few years (and discussed in subsequent chapters) strongly suggests that the consciousness experienced by all living systems is inextricably tied to a mysterious property of the quantum world called *nonlocality*.

Quantum mechanics and general relativity have produced the best understanding we have to date of atomic-scale and cosmic-scale processes of the universe, respectively, yet there are particular conflicts between them, as they are each incomplete models. From the time of Descartes' dualism, through the later philosophy of materialism, consciousness wasn't a subject of interest in mainstream science; real interest wasn't generated until the final decade of the 20th century. However, when the theory of quantum mechanics was solidly framed in 1927, it was becoming obvious that consciousness could no longer be ignored, as it seemed to possess some of the fundamental pieces to the puzzle that physicists and cosmologists themselves were trying to fit together. The struggle of individual scientists adapting their views from that of Newtonian reality to that of quantum reality was not unlike the difficulty each of us encounters in adapting our worldview to the rapid changes of the last 100 years or so. The drama always begins when we realize that something about our world is different than we previously thought, and cannot be explained with existing knowledge.

Scientists of the 19th century believed (but were absolutely convinced they knew) that the tangible substances of nature, such as rocks and trees, possessed fundamental physical characteristics measurable and completely described in terms of length, mass, chemical content, and the like. Measures related to the appearance of things. Energy was not a fundamental property, but one mathematically derived by describing natural bodies in motion. The story of quantum mechanics (and to a lesser extent, relativity) is the story of discovering that matter in motion does not only possess energy, but rather that matter *is* energy-solidified energy. In a sense, matter is the death of free energy, being one of the ways it transforms into physical reality (just as matter can be transformed back into energy). When this idea was brought into the picture, our earlier understanding and thought structures about matter were turned upside down. Atomic matter is not like tiny ping-pong balls, but is rather a continuous flux of energy combining and splitting. It only appears on our scale as solid matter.

Recognizing that energy is more fundamental than its appearance as just an attribute of matter has been an agonizing process that is only now finding its way into public awareness. The appearance of things always seems more "real" than the invisible world from which they emerge. Before we are finished with discovery altogether, we will find that "knowing" is also not secondary, but results from a fundamental property of nature. First, however, it is important to understand what "knowing" is.

Since the time of Newton, scientists and mathematicians have not looked seriously at "mind," relegating it to the realm of theology. They came to believe nature was embedded with mathematics and the laws of physics—absolute rules of order waiting to be discovered. Mind came to be considered a secondary phenomenon that evolved and eventually allowed humans to discover the absolutes. In due course, the complexities of the brain could be understood, and that would solve the riddle of mind. The mystic, on the other hand, has always believed that consciousness is fundamental, and that internal visions, voices, and insights are absolutes to be taken literally—messages from The Great Beyond. Their meanings were obvious. As a result of this historical, though artificial division of viewpoints, beliefs about the nature of the universe have become more entangled, complex, and torturous to reconcile, though each has certain evidence to support the viewpoint. Paradoxes abound, with the entire edifice of our knowledge resting upon two fractured, incomplete, competitive, and teetering foundations—physical measurement and mystical insight—largely because there has not been an acceptable model for mind and consciousness.

Today mathematical models of reality require a multidimensional universe and an ineffable quantum reality in order to make a strong, but less than complete picture. Interpretations of quantum mathematics by the most eminent of scientists require a phalanx of observers to "collapse the wave function" of the universe to its existing state. Mystical models require a spirit world, a hierarchy of angels, deities, and demigods ascending to heaven, and a progression of levels descending to hell such as that represented in Dante's *Divine Comedy*. Both scientific and religious models seem a bit strained, having to make outrageous claims to account for the mysteries of our world. As I pondered this, it reminded me of what would have happened had the makers of our *Apollo* spacecraft not consulted each other during design. Nothing would have fit when they finally assembled the various components. Science and religion have a similar problem with regard to the interface between them.

But there are alternative models of reality, and it wasn't long after my return from the moon that I began to actively assemble my own. I saw that if we allow certain mind attributes to be fundamental to the processes of the universe, with mathematics as creation of mind—a language—and if the mystic allows that visions, voices, and insight are simply information the mind needs to interpret, then the artificial, but historical division between mind and matter suddenly collapses, and the most perplexing paradox in both camps evaporates; the dualism is resolved. The universe begins to look again like the universe we inhabit. The problem is one of giving interpretation, meaning, to information. It is the story of quantum mechanics that serves to illuminate these issues.

Centuries ago, long before the days we now associate with the genesis of quantum physics, a great debate was taking shape. In the time of Galileo, astronomers who had fashioned glass lenses from sand and peered at the world around them saw how the instruments were capable of shaping light and magnifying objects. The question that troubled these first astronomers and early thinkers was that light seemed most easily described mathematically as a wave. However, since the later Greek thinkers, it was believed to be corpuscular or particle-like, and from Galileo to Newton onward for three centuries, the argument persisted. One side or the other would prevail, but only for a time, as new discoveries were made, lending credence to the opposing camp. Then by 1880, James Clerk Maxwell and his electromagnetic theory seemed to definitively settle the issue. His theory claimed that light was most certainly wavelike. Of course, this too passed.

With the discovery of the photo-electric effect by Philipp von Lenard in 1902, and the solution to quantized heat radiation by Max Planck, the pendulous answer again began to swing toward Newton and those who believed light was a particle. It was Einstein, still an unknown in 1905, who put the wave and particle ideas together, with the mathematics to show that light came in little packets of energy, subsequently called photons, each carrying a quantum of energy proportional to the frequency of the light. Light, and all radiation, clearly had both wave *and* particle characteristics. Einstein had cobbled together concepts previously considered separate.

A few years later, Louis de Broglie asserted that not only light, but all matter possessed both wave and particle properties at the subatomic level. That is to say, he brought into question the fundamental way matter was believed to exist. Atoms were not like little ping-pong balls after all. This brought us to the strange new world of quantum physics, an eerie world so baffling and mysterious that both Einstein and Planck had difficulty accepting it until the very end of their lives.

The dual nature of matter as both particle and wave is the foundation of quantum physics; we now call it the wave/particle duality.[1] Since the days of Newton, waves and particles have been given precisely definable and measurable attributes, though the definiteness of the attributes of each are quite different. Particles may be said to have a definite position, mass, velocity, and spin. Their momentum and energy are attributes that can be computed. However, waves have no mass, no definite position, nor spin. Waves can also overlap constructively or destructively (in other words, coexist at the same location)—matter cannot. But waves do have polarization, energy proportional to frequency, and a constant velocity (the speed of light in free space). These different attributes and their measures were and are a source of consternation to many physicists. How can such different concepts be brought together?

Einstein resolved one critical issue even before the ascension of quantum theory by postulating that the energy equivalent of matter at rest was the product of its mass and the velocity of light squared, expressed in his famous equation, $E=MC^2$. This equation helped relate the energy of waves to that of particles: Energy (waves) equals mass (particles) multiplied by the square of the speed of light. More conundrums arose, however, when physicists looked closely at how to manage the differences in the other measurable attributes. These problems resulted in several fundamental principles and two major paradoxes that still create confusion nearly a century later.

Two of the principles are those of *complementarity* and *uncertainty*. The complementarity principle states that particle and wave characteristics are not mutually exclusive, but complementary to each other, in that both are required to measure basic nature. Complementarity specifies which attributes of matter may be measured together, and which may not. It would be like describing apples and pitchforks with the same limited set of adjectives.

Let's assume, for example, that apples are primarily measured by their nutritional content, and pitchforks by the amount of hay they can lift. We know that apples and pitchforks both have measurable size and weight. Pitchforks have some of the elements that make up nutrition, but not many. An apple might lift a few straws of hay, but not many. Therefore, when we discuss apples, hay-lifting capacity is not fundamental, and when discussing pitchforks, nutrition is not fundamental. If, instead of particles and waves, scientists suddenly discovered that beneath the level of our visibility all nature seemed to be made up of tiny apples *and* pitchforks, we would have a problem. Our instruments are calibrated to measure weight, size, nutritional content, and hay-carrying capacity. The instruments do not measure apples or pitchforks directly, but only some of their measurable attributes.

However, the picture we carry in our minds would most likely consist of tiny versions of the apples and pitchforks we observe in daily life. If, in an experiment, we detect a little nutritional value, is the instrument detecting the apple part of nature or a little residual pitchfork? Or does it matter at all except in terms of our visual image of how we think things really are? In this way nutritional content and hay-lifting capacity must be considered complementary attributes, so long as we are tied to the notion that apples and pitchforks are what we're looking for.

The apple and pitchfork example was not chosen haphazardly. Though particles and waves seem to be a more rational basis for describing the fundamentals of nature, all such descriptions from the macro-scale world of everyday existence are just keys we have chosen in order to represent our map of reality. The underlying basis of our existence certainly isn't little apples and pitchforks, but little particles and waves may not be the best description either.

This idea of the complementary appearance of things finds its roots in the process of knowing, not in the process of what it is that exists. In order to know about something, we must first label what it is we observe, and then assign qualitative attributes such as size, weight, color, beauty, what have you, and finally ways to describe how much of the attribute is present. In science, "how much" is defined by numerical measure. In ordinary language we use adverbs. We can unequivocally state that mathematics is a linguistic creation of mind, not an intrinsic characteristic of nature, because it depends upon how we assign labels to nature, and then quantify those labels. Waves, particles, apples, and pitchforks are the blue and red lines on the map, not the roads and highways themselves. And this distinction is fundamental in resolving questions of mind/matter interaction. The map is not the territory.

The concept of waves and particles, and their associated measures, is the best metaphor from the macro-world that we currently have for basic physical reality, because energy, which is common to both, is not something we can readily picture.[2] Early in the 20th century scientists understood even less about how the brain/mind assimilates information than we do today; that is to say, how it is we know what we know. The Cartesian conclusion that the role of mind in physical measurement was unimportant, dominated science until late in the 20th century. But the framers of quantum theory recognized that the observer's mind was somehow important in the observation, and that there were hidden mental traps lurking in the invisible quantum world. They consistently warned us not to create and rely upon visual images of the underlying quantum reality, but rather to place faith only in the measurements derived from experiment. This became known as the Copenhagen interpretation of quantum physics. As many scientists of the period were dogmatic materialists, they were skeptical because quantum theory seemed to enter the forbidden mystical domain of theology—which indeed it was. They preferred a physics that was mechanistic. But how else can questions about consciousness be addressed, and the Cartesian dualism resolved, except by addressing the issue of mind? While the Copenhagen interpretation helped avoid one mental trap, it set up another involving the meaning of measurements, which still creates controversy.

The uncertainty principle specifies that complementary variables may not be measured with equal precision simultaneously. It also defines the limit of precision (thus the limit of our knowing) in the measurement. The newcomer to physics often chafes at the idea that there is a limit to the precision of a measurement. He or she wants to believe that with better instruments or technique the desired precision can be obtained. But this is just the natural conclusion of the erroneous belief that waves and particles represent distinctly different entities, when in fact they do not. Better instruments will never change that.

Problems also arise when we consider what is required to measure at this tiny scale. We must keep in mind that it takes at least a single photon of light to illuminate and thereby measure a particle. And to know in science is to measure. The energy imparted to the particle through the act of observing changes the attributes of the particle and makes its future state uncertain, a fact that must be considered in the measurement process. Historically, some have reasoned that the attempt to know is not only the cause of the uncertainty, but that it is, therefore, the process of knowing that influences the quantum processes of nature. But neither idea is correct. It is with this interpretation, however, that the Cartesian edifice of separateness of mind and matter begins to crumble.

The fuzzy line between existence and knowing was destined to become even more confused by this interpretation, and led to more paradoxes. Some modern textbooks still produce contradictory viewpoints that reflect the desperation of those attempting to interpret quantum theory. It took 70 years of experimental evidence and an understanding of how the mind manages information to bring some clarity to the issue. The account I am presenting here relies only upon analysis of mind processes, consciousness, and information. Yet it is corroborated by the experimental evidence from physics. It has occurred to me that the principles of complementarity and uncertainty apply equally well and in the same manner to the two means by which each of us observes: the sensory, or outer experience; and the inner, subjective experience. These two modes of observation are complementary in that both are required to complete our picture of reality. Yet they have different characteristics, and are not equally valid for all observations. The outer experience reaches its zenith in the scientific method, and is characterized by specificity, precision, and detail, and most often uses the language of mathematics—it is understood intellectually. The highest expression of the subjective experience is in the ineffable mystical insight. It is characterized by holistic patterns, and lack of precision, but deep feelings of certainty. By necessity, it must be expressed through metaphor, analogy, archetypal imagery, and the arts—its effect is emotional. The notion that science and religion are mutually exclusive expressions because they each emphasize a different mode of observation is no more valid than the idea that waves and particles are separate physical things. In both cases complementarity and uncertainty prevail, and in both cases it is mind that gives meaning to experience. Dogmatic acceptance of the Cartesian duality, particularly in the Western mind, has created an artificial barrier. Although science arose and has flourished for nearly four centuries as a result, a new synthesis is now required in order to proceed. *Objective observation* is not, strictly speaking, possible, as all observation is inherently a subjective experience.

The formulation of the principles, and the mathematics of complementarity and uncertainty for quantum observations, led to the notion that between the measurements of wave/particle attributes we cannot know precisely what the attributes are. All we can know is that the wave/particle obeys certain rules, such as the conservation of energy and the conservation of momentum. The value of the other attributes between measurements came to be specified as a probability, not as a knowable value. Only when measurements are made can one "know."

Some theorists go so far as to say that nothing actually exists between measurements except a probability. With this interpretation, the territory has disappeared altogether, and just the map remains. The equivalent mistake would be to assume that just because a pilot is flying through clouds and cannot see the ground, mountains don't exist.

Although there are two major methods for computing and keeping track of the probable attributes of wave/particles, the one most often used by practicing physicists is called *Schrödinger-Wave Equation*, or *Wave Function*. This wave equation has terms representing the attributes to be considered, and a particular property: if certain terms of the equation are squared, the terms represent the probability that the measurable attributes will have a certain value. The wave function is considered in some quarters to be that which moves through space-time between measurements, rather than the wave/particle itself.

For example, one can compute that for a sealed jar of water, there is a calculable probability that a molecule of water will escape through the glass walls, and its position will actually be outside the jar. This is why hazardous substances, such as radioactive materials, often require thick, lead-lined containers—to reduce the probability that much will escape. Because all matter has the properties of quantum duality, there is also a calculable probability that the mate you left behind at home in Los Angeles will bump into you on the streets of New York while you are there, even though that probability may be very small. But until the event happens—that is, until the observation (measurement) is actually made—only its probable values can be known. When the measurement is made, the value of the attribute is then certain, and the wave equation with its probabilities is said to "collapse" to an actual event with probability equal to 1.

In classical Newtonian physics it was believed that if you knew the position and velocity of all particles in the universe at one time, it would be possible to compute their future positions, and make predictions for all time. Therefore, it seemed possible to one day predict the future with unflinching accuracy, because it was believed that we live in a universe governed by the elegantly simple, deterministic laws of Newtonian physics. Events, our lives, the weather, the progression of the universe itself, would one day be subject to definitive forecast. But the existence of wave/particles that can have any number of probable values for their position and velocity in space-time between measurements causes the universe to be far more interesting, mysterious, and disturbing than one that is elegantly deterministic—and presumably less knowable. Yet a universe of probabilities still falls short of accounting for intentionality and choices that are made every minute.

Einstein, at once unconvinced of the shortcomings of classical determinism and disturbed by the counterintuitive idea that one could say nothing about the underlying structure and movement of matter in the interval of space-time between measurements, devised various "thought experiments" to probe the validity of quantum theory.

Exchanges between Einstein and Neils Bohr through the course of several decades proved the value of mentally dissecting experiments to discover truths and flaws. Because events at the atomic scale were too small to observe directly, thought and imagination were needed to create and think through both physical and thought experiments. As a result, the subtle line between external and internal reality, between physical experiments and thought experiments, began to blur even more. It has taken years for technology to catch up so that the subtleties of thought experiments could be physically tested. One such thought experiment has particular relevance to our story, as it illustrates the "supernatural" ways of nature.

Imagine that protons, ejected from a source, go shooting off in different directions, tumbling and twisting as they go. The probable values of the attributes of each separate proton may be described by a separate Schrödinger wave equation. The values of the attributes of the two particles must remain correlated to each other, because they came from the same process and must obey the conservation laws, even though each at any moment could display a range of values for any particular attribute.

Were one particle to be captured and measured in Princeton, New Jersey, for example, the other might at the same moment be in Bangkok. Values of the Princeton particle are suddenly known, and its wave equation collapses to known values. The Bangkok particle, could it also be captured and measured simultaneously, would have to display the appropriate values for its measured attributes because energy and momentum must be conserved, and the quantum attributes linked. These are fundamental laws of both quantum and Newtonian physics. What no one understood was how the Bangkok particle would *instantly* know that its partner has been captured, and its wave equation collapsed to correct specific values without violating the proven relativistic notion that signals cannot travel between Princeton and Bangkok faster than the speed of light.

This conundrum became known as the Einstein, Podolsky, Rosen (EPR) Paradox, and it was considered not only critical but downright spooky. Was determinism and its picture of underlying physical reality really lost, or was quantum theory merely an illusion, a phantom result of a flawed intellectual game? It seemed that nature could act "supernaturally" at the subatomic level. Particles were capable of communicating across oceans instantaneously—that is, *nonlocally*. And I had my own reasons for wanting a satisfactory answer, as the behavior of these particles raised many of the same issues as the abilities displayed by Norbu Chen and Uri Geller.

16

Author's note: The following two chapters present a somewhat detailed examination of quantum mechanical interpretations and paradoxes that have resulted from lack of understanding of mind processes. They were written for those with a knowledge of or curiosity about quantum paradox. The casual reader is invited to skip to Chapter 18.

We now know Einstein was wrong in the extent of his trust in classical physics to explain certain phenomena. The probability rules of the wave equation of which Einstein was skeptical do indeed help us map quantum processes. The EPR Paradox was formulated into a testable statistical theorem called Bell's Inequality in 1964, but its results were not demonstrated by decisive experimental evidence until 1982. A physicist at the University of Paris by the name of Alain Aspect led a team that demonstrated that the particles (Princeton and Bangkok) faithfully maintained the proper correlation when the wave function was collapsed by independent measurements. Aspect used the polarization of photons instead of protons, but the results would turn out the same.

The popular interpretation of their findings is that the relativistic speed of light is not violated; that is, information is not transferred from one particle to the other, but rather the wave aspects of the particles are in some way interconnected nonlocally, and "resonate" so as to maintain the correlation of their characteristics. They do not behave as particles at all, but rather as fields, filling all space, orchestrated and mediated in their properties by a mechanism not yet understood. Quantum theory does not and perhaps cannot, in its current form, address how the properties are orchestrated and mediated nonlocally to maintain their correlation. Additional theory is needed to understand nonlocality. But one alternative explanation, that information is transmitted faster than the speed of light in space-time, and that particles each "know" what the other is doing, is even more problematic. In any event, the experiment conclusively demonstrated that matter and information have attributes that remain correlated nonlocally. It's this mysterious *nonlocality* that brings new insight into a number of problems, including those of many enigmatic, subjective attributes of consciousness. This paradox also demonstrated that the methods of science are sufficiently powerful to uncover the flawed assumptions in the Newtonian and Cartesian thought structures.

Because the rules of quantum theory are supposed to apply to all matter, not just subatomic matter, by extension of this ubiquitous, interconnected "resonance," it suggests that all nature is in some sense wavelike, "fieldlike," and "mindlike," in a way that isn't yet fully understood. The experimental results imply analogies to the silly superstitions of some medieval (and modern) mystics, and are somewhat akin to the Platonic idea of the real, unmanifest world as made up of ideas and perfect form. Resonance and nonlocality are fundamental clues to all psychic functioning, and are attributes of the newly discovered *quantum hologram*, which is discussed in a subsequent chapter.

The second major paradox arising early in the history of quantum theory still hasn't been answered to the satisfaction of most physicists. It's another thought experiment called the Schrödinger's Cat Paradox. Erwin Schrödinger formulated the wave equation that bears his name, and the paradox derives from use of the wave equation. The experiment is especially relevant to our understanding of consciousness, as it arises directly from the entanglement of existence and knowing; from confusing the map with the territory.

By following the Einstein-Bohr procedure of minutely dissecting thought experiments, one can reveal hidden traps in the vast labyrinth that produced this paradox. The traps all have to do with how we organize information in our mind/brain and give it meaning, and the subtle dividing line between internal and external events. Because the Schrödinger wave equation allows one to compute the probable values of something's attributes (generally referred to as its *state*), this thought experiment concerns the state of a cat. It's a peculiar imaginary situation, but one that's immensely important.

Suppose a cat is placed in a box that contains a radioactive substance, a flask of poison, and a trigger mechanism. This odd collection is arranged in such a way that the radioactive decay of the substance creates a 50/50 probability that the vial of poison is broken, and if it is, the cat is instantly killed. Particles do not emerge from decaying radioactive materials on a uniform time scale, but randomly, and such that half of the total emitted particles will escape in a period called the half-life of the radioactive substance. Half of the *remaining* half will escape in the second half-life interval, and so on. The apparatus is arranged so that after a fixed interval of time, the probability that the vial is broken is precisely one-half. Then the experiment is stopped. Thus, a question arises: Is the cat dead or alive?

The cat's wave function no longer changes with time, because the experiment is stopped and the wave function remains in the state that gives a 50/50 chance that the cat is alive. But the cat cannot be in suspended animation, half-alive or half-dead—only completely dead or completely alive. Or can it? The experimenter does not know the answer until he has looked in the box and observed (measured) the cat. The common sense Newtonian answer would be that the cat is completely dead or completely alive and the experimenter's knowledge has nothing to do with it.

But quantum mathematics has proven time and again that nature defies both common sense and the Newtonian answer. That's what makes this experiment so interesting. Physicists must trust measurements and mathematics because only experimental verification can validate theory. The Copenhagen interpretation of quantum theory unequivocally requires that the state of a quantum object can only be described as probabilities between measurements (observations), which will collapse the wave function to an actual state with probability equal to 1. But what qualifies as an observation or measurement that will collapse the wave function? One interpretation of the meaning of quantum mechanics insists that the superposition of the probability waves, one for a live cat plus one for a dead cat, is the only reality. The collapse of the wave function clearly has occurred once the experimenter knows the state of the cat. But is it the observer's knowing that is required as executioner? Were we to substitute a philosopher for the cat, would the answer be different?

Presumably a philosopher could know and signal his state to the experimenter. But science has not recognized subjective data as valid—particularly that of philosophers and mystics. Therefore, the subjective self-assessment of the philosopher as to his own state of aliveness may not be acceptable to the experimenter. But we have to question whether the experimenter's distrust of the philosopher can prevent the collapse of the philosopher's wave function.[1]

Such thought experiments are not in and of themselves absurdities; they push our thinking to the limits so that understanding of flaws and new insights may emerge. They bring to light the subtleties that must be seriously considered when attempting to understand the strange world of quantum mechanics—indeed, to understand the universe at all. The Schrödinger's Cat Paradox brings into question the role of *knowing* in making observations, and thus raises the question of whether mere knowing can alter the physical world around us, and if so, how.

The interpretations of the cat paradox that allow wave function probabilities to be construed as actual reality led to a swamp of absurdities when viewed from a broader philosophical viewpoint. Three generations of physicists have used these interpretations with few questions, because quantum mathematics works like a cookbook recipe when applied to practical problems. Philosophy is not required. But the map is not the territory, and mountains still lie beneath the clouds even though the pilot can't see them. It has taken several decades of experimental evidence to demonstrate what is required to restore a degree of coherence to the meaning of quantum mechanics. The fundamental message here is that mathematical descriptions are not embedded in nature, but are created by *mind*. The differences between existence and knowing must be carefully observed when considering the larger viewpoint.

In the early days when scientific tools were relatively crude and simple, scientific observation and the measurement of results were a single event. Observation and measurement were, in a broad sense, synonymous. But today that's no longer true.

Take for instance the ancient philosophical question of whether a tree falling in the forest makes a sound if there is no one there to hear it. The mind game has a new spin today. The question must be answered differently with the availability of modern technology, particularly with the advent of remote recording devices. The falling tree creates a pressure wave, which a pressure-sensitive instrument can record if tuned to the proper frequencies. But that is neither hearing nor sound. Hearing and sound relate to the *subjective* experience of a living organism.

However, according to quantum theory, if neither the person nor the device were there, one could not know or say what is happening with the tree and its pressure wave. By one radical interpretation, the tree doesn't even *exist* unless it's being observed. The beauty of the scientific method is that if we were to study many falling trees we might gather subtle clues as to what happens when trees fall, and then *deduce* what happens when the tree falls in the middle of the forest without an observer or a recording device. But is deducing the same as knowing? It certainly isn't the same as measuring.

Einstein criticized quantum theory when he stated that God doesn't play dice with the universe. In the metaphorical sense of the statement, he was wrong: The universe does indeed obey the rules of quantum probability. But with the knowledge we have today of the evolution of the universe, its species, language, and the thought process, it can be said that he was absolutely correct in the literal sense. God does not in fact play dice with the universe, because dice, the rules of probability, and even the meaning and nature of God are human creations of the last several millennia, and the universe is apparently several million times older than that. The subtle nuances of words, construed literally or metaphorically, can shift the truth of knowing.

Whenever trees have fallen in the past, they likely behaved just as they had before humans made an issue of it by asking annoying questions. Likewise, the universal processes proceeded to unfold as they did before quantum mechanics, Newtonian mathematics, and telescopes— long before humans began to second-guess the processes and create these "maps." The territory they represent seems to have existed long before our maps and the anthropoid mentality that created them.

The anthropic consciousness and mentality we experience today certainly seems important to the future evolution of the universe. We have radically altered the face of the planet, particularly in recent years, not only by our very existence, but also because of intentional choices. If consciousness is important to the universe, what then is its exact nature, and how did it manifest before we humans evolved to the state of awakened awareness we currently enjoy? An investigation of quantum mechanics with consciousness in the picture begins to yield some interesting and useful answers.

The now-classic wave/particle demonstration, called the *double-slit experiment*, has undergone exotic refinements through the years to help us better understand this strange phenomenon, and to shed light on the several paradoxes of quantum theory. An example of the experiment consists of a source of light, a photosensitive screen to capture the light, and a device that can split the light beam into two beams at the will of the experimenter before it reaches the recording device.

In principle this is a simple experiment. If only one beam of light is used, all the photons cluster in a blob around the center of the recording screen, just like a flashlight beam projected on a wall. If the beam is split into two beams by a beam splitter and then refocused, an interference pattern showing alternate light and dark areas will appear on the screen. All wave equations predict this result, but particle equations predict two independent blobs of light—not an interference pattern. This was the original proof of wave/particle duality.

All refinements of the double-slit experiment have been made with one or both of two primary goals in mind: The first goal has been to catch the wave/particle behaving as a Newtonian particle between measurements, thereby showing that classical ideas and pictures of reality prevail. This, of course, would deny the distasteful uncertainty principle, making the proponents of classical physics very happy. The second goal has been to discover how, what, or when the wave/particles know. But even though the wave/particles have remained enigmatic, a few telling facts have been gleaned.

The various versions of the double-slit experiment have consumed more than half a century of finagling; they even include experiments that fire photons at such a slow rate that they can be observed individually. Another experiment, called the *delayed choice*, allows the apparatus to change after the photons have presumably made a choice as to which path they will follow. In the *quantum eraser* experiment, data is recorded by a computer and can be selectively viewed in parts or as a whole to see which photons used which path.

The results that emerge from these experiments are that the wave/particles always "know" whether it's a single- or double-slit experiment being conducted, and they behave accordingly at the instant they are captured on the collector screen. It doesn't matter when the experimenter chooses to change the experiment, when to look at the results of the experiment, or whether to look at all or just some of the data: The wave/particles always behave appropriately to either a one- or two-path experiment when they relinquish their energy to the collector. What this means is that time doesn't matter to the wave/particles. Nor does it matter what the experimenter knows; it only matters if the results are knowable; that is to say, if they are recorded on a device, as when a tree falls in the forest. But the particles always maintain the proper correlation.

The best answer to these mysteries seems to emerge from EPR Paradox and the Aspect experiment (Princeton and Bangkok). The attributes (states) of the photons must remain nonlocally correlated to the values of their wave equations wherever they go until captured. Though tempting, it's misleading to apply anthropic terms such as *knowing* to particles, as they are just in resonance and maintaining correlation. Moreover, larger particles, which can betray their presence by emitting radiation, have now been put through double-slit experiments, and it's possible to detect the passage of the particle without disturbing its path. The results are the same, but what's most important is that this demonstrates that the wave/particle still retains its "existence" as a particle even when behaving as a wave and going down two paths simultaneously: It is not just a probability wave; it has real existence. We just can't say what the values of the attributes are without measurement. But they actually exist apart from our knowing about them. The wave function of a particular wave/particle collapses at the instant the energy of the wave/particle is delivered to the recording device and its position becomes knowable by the experimenter.

What this tells us is that the wave/particles respond to the process they are undergoing at the behest of the experimenter, whatever that process is. Time, in its common usage, is not important to them. Time is just our measure of the change of states involved in a process. Time is about our knowing, and a measure of the speed of a process. It does not have independent existence in nature. Changing the process during the experiment causes immediate adaptation by the wave/particles to maintain nonlocal resonance with all parts of the setup, which are instantly correlated throughout the experiment. When the wave/particle gives up its energy to the collector, it ceases to be that wave/particle.

If we apply this idea to the Schrödinger Cat Paradox, what do we have? Replace the cat with a poison-molecule detector—one which records the molecules' time of impact. If the detector registers a poison molecule by the time the probability is 50/50 that the cat is dead, then we can deduce that the cat would have died. *When* the experimenter reads the recorder and discovers the result is irrelevant. Were we to leave the poison-molecule detector in place and add the cat, a philosopher, or a scientist, the only difference would be the addition of a wave function representing a conscious, sentient, quantum entity. The conclusion: flawed because it unequivocally requires that the state of a quantum object can only be described as probabilities between measurements (observations). The modern understanding that all physical objects of every scale size possess quantum attributes (a quantum hologram) nullifies that earlier interpretation.

17

Not long into my research I began to ponder what it was about a living being that changes the wave equation we would use to represent it in an experiment. Would a simple rock's wave equation remain unchanged, while a cat's would not? Representing the effects of consciousness and sentience with a wave equation first requires one to be able to specify and measure consciousness and sentience.

Asking how and when the experimenter's state of knowing collapses the wave function, without first discovering how a rock, a cat, or a philosopher knows anything at all, would not likely yield any useful answers. The major problem lies in the confusion of maps and territories. The maps are the knowing, the images, the thinking, the beliefs, the measurements that we carry around in our heads. They are indeed the only information about reality of which we are aware. The territory is the world itself. At any moment our *conscious awareness* is not accessing all the information available to the brain/mind, and the brain/mind is not accessing all information available in the world around us. If it were, then our map could precisely depict the territory; then we could really "know" reality. Yet in an evolving universe it must be an ongoing process of learning. We humans have a tendency to believe that whatever we have in our heads is Reality and Truth, at least for the most part, because it is the only map in our possession, and we don't intentionally riddle it with errors.

But I don't know anyone who has found life and reality to be always as he or she has interpreted it. Daily life is always correcting us, challenging our beliefs from moment to moment. What is in our minds is not a perfect reality, but rather a shadow world based on incomplete information and assumptions about whatever reality is. Although our only map is incomplete, it still governs our behavior and thinking as we "create" our personal reality. It seems accurate until life's experiences, usually unpleasant ones, convince us to add new information and reinterpret the meaning. Misunderstanding this process by which we acquire, interpret, and know is the primary cause of all human folly. Certainly the scientists attempting to understand general relativity and quantum physics (and other subjects) didn't have a proper understanding of this process for most of the 20th century. Lack of a proper model for the interactive processes of existence and knowing has led to inappropriate retention of certain Newtonian absolutes and beliefs disguised as knowledge. The Newtonian model is correct, however, with regard to there being existence apart from our knowing about it.

But does the experimenter's state of knowing influence the outcome of events? Were the wave function to accurately represent the experiment, and the experimenter's state of knowing actually to collapse the wave function, then the experimenter's choice to know and when to know the outcome of the experiment would force the outcome to manifest itself in reality. This is just a fine hair short of the Idealist or Platonic position, which holds that the deepest reality lies in the mind and manifests itself through thought. The extreme position of the Idealist would be that the experimenter not only *chose* to know the outcome, but *created* the entire experiment and its outcome only in his own mind. Physicality follows the mental image. A thought experiment and its physical enactment are one and the same. In other words, the cat was either dead or alive at the experimenter's whim—were the experiment actually performed.

The first major flaw in this line of thought is the failure to distinguish the differences in the way we experience external and mental reality. Our senses pick up only a small part of our external reality and augment that information with our prior experiences (memory) so that our mental reality is, at best, only a poor representation (perception) of external reality. So it's enough to say that normal humans experience each differently. Another important subtlety involves the differences between the mental states of being aware, knowing, and intending. Being aware and knowing have important differences: knowing requires attaching a meaning to information based on prior experience that conveys a sense of certainty, whereas being aware does not. It is actually being aware of whether the cat died that is at issue. Would being aware of the cat's state of aliveness influence that state? Of course not.

Causality is still alive and well in our universe, though both relativity and quantum theory call it into question. However, causality may now be understood quite differently than in older thought structures. Being aware only allows the screen of one's conscious awareness to fill with information. It causes nothing but a few million neurons in the brain to snap to attention. None of the traditional thought structures, nor modern studies for that matter, suggest that merely being aware causes external action. The experimenter's awareness of the cat's state when opening the box may have caused the experimenter's internal map of the wave equation to collapse and coincide with the alive or dead state. But the experimenter's wave equation is only being updated by external information. Garnering that information does not, by itself, influence the cat's health, as the pattern of energy flows from cat to observer. The experimenter might have become aware through nonlocal means (intuitively), but that still isn't causal. Were the experimenter psycho-kinetically active, we might possibly influence the apparatus intentionally. Therefore, simply being aware couldn't kill the cat, and knowing couldn't kill the cat. But intending to might.

In any of its various forms, scientists usually disdain the Idealist position, which allows the mind to be causal. Yet Schrodinger's cat has been in a box with a vial of poison, waiting 75 years for some experimenter to decide whether or not looking in the box created the coup de grace—whether or not the probability wave is the same as physical reality. It has doubtless died of old age by now, regardless of how we think about it. Physical death, it seems, is inevitable.

An entire body of interpretation of quantum physics has grown from the Schrödinger's Cat Paradox. The conundrums sprouted from the perplexity of the issue and the lack of understanding as to how mind and body interact (Newton erroneously assumed they don't interact). An interpretation of quantum mechanics exists that is really an attempt to get around the unsettling indeterminacy of quantum theory and the lack of a "picture" for the underlying reality. This is the Many Worlds interpretation, and it proposes that all of the probabilities in the wave equations are real. By simply choosing to open the box and discovering the cat dead or alive, the experimenter causes the universe to divide into two universes: one where everything is the same, except the experimenter

From Outer Space to Inner Space

discovers a dead cat, and another in which she finds the cat alive. Here the map *is* actually the territory.

In this interpretation, each choice we make branches the universe into the many probabilities available. The one we find ourselves in corresponds to the choice we made, but the others are presumably equally real. It's just that we can't be aware of them, because they are orthogonal (at a 90° angle) to our own. This interpretation by physicists is a permutation of the extreme position of the Idealist model—though some physicists have trouble admitting as much.

Many Worlds is a flawed concept that unfortunately has yet to be falsified since it was proposed in the 1960s. I find it both amusing and absurd that physicists will believe that even casual choices can create entire universes we can't see or verify, and must violate conservation laws. Yet they can't see their way clear to acknowledge psychic resonance or intentional psychokinesis, which can be observed and verified and *don't* violate conservation laws, as they utilize nonlocal resonance, entanglement, and coherence.

If the Many Worlds interpretation were true, we would never find ourselves in a universe that disagreed with us, because we would always choose what we believe or what we want to believe, even at the subconscious level of our understanding. Humans will rarely choose pain unless they believe it will lead to pleasure. Therefore, if by conscious choice we could branch our universe and remain in a pleasant one, we would have all doubtless done so by now. In such a universe, scientists would soon find all their own theories and experiments succeeded, while those of their critics blew up in their faces. In any form of the Idealist thought, it doesn't seem that nature could invalidate our cherished ideas, yet nature often does just that. In a Many Worlds universe, most of us would find ourselves in Neanderthal heaven, since we could have avoided the responsibility and drudgery inherent in evolving a civilization, choosing instead to discover lush, green fields and easy hunting grounds.

The mystical version of Many Worlds is a bit more difficult. Here, one must put aside conscious, thinking choice and choose with the heart (the subconscious). In Christianity one must choose with the heart and have faith. In most East Asian religions, one must keep practicing the choice until union with the godhead (heaven, nirvana, samadhi) is achieved in this life or the next. I've personally worked closely with several of these ideas, and my personal, internal world has steadily grown more pleasant. I have not, however, noticed that the world I live in has improved correspondingly, though I've consistently wished it so most of my life. Merely visualizing world peace does not by itself beget world peace.

Einstein quipped that the moon doesn't go away when we close our eyes. Only if the Idealist position were in fact true would it be the case that physical objects in the macro-world vanish when we cease to look at them or don't have them in mind. Only if the moon were just a probability wave in the mind would this not be an absurdity. In fact, the wave equation for macroscale objects emphasizes the particle aspect and diminishes the wave aspect of such objects so that the probability of the moon being anywhere except where Newtonian equations predict it to be is vanishingly small over any finite time period. Therefore, we don't really have to worry that the moon might vanish the next time we close our eyes. However, the fact that it could, even with a small probability, apparently provides sufficient latitude for the intent of Uri Geller to affect recovery of Edgar Mitchell's tie pins.

18

Information was first defined in scientific terms by Norbert Weiner of MIT, the father of cybernetics, circa 1942. He gave it an elegantly simple definition: the numerical equal to the negative of entropy. In a famous paper published in 1948, James Shannon of Bell Laboratories initiated information theory from which all modern communications techniques are derived. Engineers and scientists agree that information is basically just a pattern of energy. What is tacitly assumed is that the meaning of the information is carried in the signal itself. In other words, any mind should interpret information the same way; if it doesn't, then it is because that particular mind is ignorant of that particular meaning. To understand how consciousness knows anything at all, we must examine this idea more deeply, as it contains a critical flaw.

Imagine a family of four: a mother, a husband who is a fireman, their child, and a mother-in-law who is deaf. One night while they are all asleep a fire engine from the husband's station roars down the street, passing their house, sirens crying through the neighborhood. The wife awakens, hears the fire truck rush by, and thinks, "Thank God, it's not our house." The husband awakens and thinks, "Sounds like ol' #4. It's probably Joe heading for Oak Street." The child awakens, frightened by the loud noise, and begins to cry. The mother-in-law is roused, but doesn't remember a thing and goes back to sleep.

Each of the three who received the information attached a different meaning to the sound. One received no conscious information at all. The usual meaning intended for fire truck sirens is, Get the hell out of the way, I'm coming through. Clearly the meaning attached to information is not discerned the same by all who receive it. If there is _intentional_ meaning in the signal, it must be placed there by some intending entity, presumably one that knows how to manage information intentionally.

Managing information intentionally is, of course, a wonderful, basic definition for intelligence. Therefore, we cannot state that the meaning of information is contained in the signal. Even if it were, there is no assurance that any receiving entity will perceive its meaning.

The meaning of a signal, therefore, must be defined as the meaning the *percipient* of the information attaches to it, which may be as varied as the number of percipients. Thus the intended meaning and the meaning the percipient attaches to it may not be one and the same. This fact is obvious when one stops to think about it, as everyone has experienced failed communication with someone. Only by some method of achieving a consensus between the sender and receiver can common meaning be assigned by the *exchange* of information.

Likewise, the patterns of energy in nature contain no discernible inherent meaning. Ancient astrologers attached a different meaning to celestial patterns than do modern astronomers. I suggest that it is a fundamental function of consciousness to assign meaning to the information that the organism perceives. Meaning is merely additional information that a conscious entity internally creates by comparing the new information with information stored in memory from previous experience. Meaning is internally generated information that connects and explains the world as we experience it. Meaning is our individual interpretation of events.

The meaning of the universe, therefore, is what the perceiving organism assigns to it. If there is inherent meaning in the universe it is because the cosmos is in some sense a conscious entity and has intended such meaning. To comprehend such meaning would require an exchange of information with the universe—which is precisely what the prayers, meditation, and rituals of mystics are designed to accomplish. Prayer is information intended to be perceived nonlocally; meditation quiets the mind and expands awareness so that nonlocal information may be perceived locally.

From the earliest days of science, even before Newton, inquisitive minds attempted to describe the natural world in terms of simple mathematical relationships and geometric shapes, believing that what they observed in the ordinary world ultimately represented a reality that could be explained in simple terms. Perhaps the idea stemmed from the Platonic notion that the real world is the world of beauty and perfect form, and the observed world a shadowy, imperfect imitation. Perhaps it arose because our thinking and knowledge had to evolve from the simple to a more complex form following the creation of spoken and written language.

Whatever its source, the human thought process would prefer the world to be simple and explainable, to fit into the neat little pigeon-holes of language. But simple, ordinary terms and linear mathematics, as it turns out, can only describe near-equilibrium processes such as water slowly warming on the stove or flowing gently down a river. They cannot describe violently boiling water or water rushing through rapids. The overall energy exchange between the sun, the Earth, and deep space is a near-equilibrium process, but the violent weather that moves across the face of the earth is not.

It wasn't until the late 1970s that a physicist by the name of Ilya Prigogine described the most pervasive and important processes in the macroscale universe as far-from-equilibrium processes. He garnered the Nobel Prize for his work on dissipative structures, which he reported in his book, *Order Out of Chaos*. Such processes require nonlinear, complex mathematics to analyze. Prigogine points out that near-equilibrium processes, which can be described by simple, linear mathematics, are quite rare. Near-equilibrium mathematics serve in the laboratory only to crudely approximate the more interesting complex processes of nature. I find it astonishing that with the simple, linear tools and analyses we've used in the classical sciences for 300 years, that we've come to understand the universe as well as we have.

Only 50 years ago, my professors in graduate school still eagerly pointed out simplifying assumptions that reduced the complexities of fluid flow problems to terms that were easier to understand and utilize. Earlier scientists could only ignore the complex natural events for lack of the tools to comprehend them. With the invention of high-speed computers that can simulate the complexities of nonlinear processes, the problems found in arcane new subjects, such as dissipative systems, chaos theory, and nonlinear optimization of systems, could be programmed, providing us with new insights into nature's more complex physical processes. That's not to say that these problems have been solved, for few nonlinear problems have neat, closed-form solutions. The computer merely provides a new way of mapping the mysterious landscape of nature by evaluating the same equation thousands of times, starting with different numbers each time.

The salient characteristic of Prigogine's discoveries about dissipative structures is that they are entropic processes, meaning they are not time-reversible. Being highly nonlinear, they are also not deterministically predictable. Examples of irreversible dissipative processes with which we are familiar in everyday life are the acts of birth, living, dying, spilling a glass of water, breaking an egg—which even the king's horses and men couldn't possibly reassemble. A familiar illustration of the non-predictability of dissipative processes is weather.

Similar to the processes of quantum mechanics, how weather patterns will likely affect a particular location is most easily described as a probability. We make jokes at the expense of weathermen, but in principle, the details of air mass movements are so complex that practical solutions are only knowable as probabilities. That's why the weatherman speaks only of a 10 or 20 percent chance of rain, or that it will be partly cloudy. Far-from-equilibrium processes contain so much energy moving so rapidly that the slightest perturbations anywhere in the system, even from a seemingly insignificant source, can cause large unpredictable consequences such as the "butterfly effect," in which it is suggested that a butterfly flapping its wings over Beijing might cause a snowstorm in New York.

Being far-from-equilibrium causes the character of the process to change unpredictably. Such transitions from predictable to unpredictable movement are called *bifurcation points*. The process at the bifurcation point branches such that it suddenly creates a whirlpool or a tornado in an air mass, for example. The fact that the macroscale universe consists largely of these entropic, nonlinear, irreversible processes, and not the deterministic, simple, time-reversible processes of classical science, helps us realize that the arrow of time only flows in one direction here in the macroworld. In other words, nature's energy only moves forward through irreversible processes. The process cannot run backwards. Nature knows nothing of time; all she follows is process as energy and matter flows in its cosmic dance. Time is an invention of mind—a part of the map, not the territory.

The consequence of this is that information, defined as a pattern of energy, cannot come from the future and be known in the present, even through nonlocal means. The future can only be predicted as a probability, because it consists of quantum entities at the microscale, and irreversible processes at the macroscale. Neither requires nature to have a notion of time; only our attempt to measure requires this. If we think of ourselves as floating in a tiny canoe through a turbulent stretch of rapids, just going with the flow, then we have a good idea as to the physics of life. We cannot predict our path from minute to minute within the river. Even though we understand the overall process of streamflow and know we'll eventually get to the ocean, we simply can't predict our exact course. With the intention of finding smooth water and a paddle in our hands, we can influence the process, but only to a certain degree. Life itself is a dissipative process.

And telling you this story, being an irreversible process, has just reached a bifurcation point. I want to discuss more about time, but also tell you about chaos. Yet I cannot discuss both subjects simultaneously, because language, too, is a linear, sequential process, which I am using to describe nonlinear processes. Therefore, I must choose.

In nature, time has only one direction. We call this the arrow of time. The origin of the arrow is still a bit elusive, but it most likely originated after the big bang, if indeed there was one, as the expanding universe began to cool.[1] The arrow of time is completely defined by the direction of irreversible processes, of which there are two in the macroscale universe: entropic and negentropic.

When Norbert Weiner defined information as the numerical negative of entropy, he had these two irreversible processes in mind. One numerically describes the process of decay, the other the process of creation. Nature, he observed, does both simultaneously: There are processes of decay toward disorder, and processes of creation and building toward greater complexity and order. Because the instant at which the big bang occurred at a tiny point represents the highest temperature and greatest pressure the universe has ever experienced, it is said by physicists to be the point of greatest order. It has been downhill from there—presumably, toward the death of the universe. Fortunately for us, the universe began to structure complex physical reality after that initial instant. The process of building ever more complex molecules, thus building a macroscale reality, is a negentropic (negative entropy; that is, an information-building) process.

That process continues to this day, creating new complex structures within the universe, while the overall universe expands outward. These are questions of cosmology, of how we and the cosmos around us came to be. But the process of the universe as understood today is that of transforming the enormous unstructured potential at the moment of the big bang, according to prevailing theory,[2] into a structured macroscale reality, which, 14 billion years later, resulted in human beings who can ask questions about that process. Perhaps we are a star's way of knowing about itself.

The arrow of time, therefore, proceeds from the big bang forward. Its direction is defined by the irreversible macroscale processes of decay and creation, both of which only proceed in one direction—forward to the future. We must emphasize the macroscale because energy, whether we observe it as wave or particle, interacts with other energy reversibly and without loss at the microscale. Subatomic particles combine, separate, and otherwise interact in a manner that may proceed in either direction. At that level of nature, the arrow of time is not defined; it has no meaning. However (and this is a big however), physicists still use equations that include time when working with subatomic quantum processes—a holdover from our classical Newtonian history.

One way of getting around the fact that time is not actually relevant in microscale processes has been the introduction of both positive- and negative-flowing time, which mathematically cancel each other out. The same procedure can be used with the equations of general relativity, in which time is irrelevant at the speed of light, but positive-flowing below the speed of light, and negative-flowing at speeds faster than light (if such a realm actually exists). This procedure gives correct mathematical answers, but is terribly messy and clearly suggests the need for a different interpretation of what's actually happening. The fact that macroscale processes are irreversible and produce entropy, and that microscale processes are reversible and nonlocal, yet produce information, suggests that a rethinking of Weiner's mathematical formulation is in order.

The arrow of time is only knowable at the level of macroscale processes in which energy is lost into the larger environment, or organizes into a more complex structure through irreversible processes. The measure of time in which we humans place such great store is only an arbitrary convention we've adopted to help measure processes from our level of observation. Time is about maps (clocks) and knowing, not about actual territories (reality). Nature follows process but knows nothing of the hour, minute, or second. In our solar system, we measure the day based upon the convention of one rotation of our Earth upon its axis. But in the world of subatomic particles, the day is irrelevant. In the Newtonian world, time is both absolute and reversible, and considered a fundamental attribute of existence. This error still clouds the larger picture.

Einstein's theory of special relativity even tells us how to transfer our local numerical measures from one moving reference frame to another. In other words, it tells us how to apply our maps to different moving points of view. But we err if we ascribe knowing about time directions and measures to levels of nature that don't carry clocks. The fuzzy dividing line between existence and knowing can again blur the distinction between external and internal. The internal reality is that with which we know, but both the external and internal are what we attempt to know *about*.

From the point of view of our internal thought processes, time can become whatever we want it to be.[3] Psychological time can go forward, backward, slow down, or speed up. We have complete latitude in our imaginations to play with it as we choose. Such thoughts can even interact with the processes of external reality through our intentions. We actually have only one time, and that is the *now*. Whatever we do with imagination, *now* is all we know. The past is but history, information stored in various forms. The future hasn't been created. Our bodies, however, respond to the diurnal cycles of energy and to our internal processes, including thought—not the clock, which represents a technology, not nature.

If we consider the results of dissipative structures on our concepts of time, then compare them to discoveries in relativity and quantum physics, we find interesting common ground. Photons and all electromagnetic waves traveling at the speed of light are in a domain where time, according to scientific theory, stops in that reference frame. Relativity gives us one interpretation: As matter approaches the speed of light, time slows to a stop, as matter requires infinite energy to accelerate. Or, said another way, the mass of an object becomes infinite as an object approaches the speed of light. Another interpretation is that time has no meaning in that reference frame, and only the wave nature of matter can be manifest. This latter interpretation is quite in keeping with the Aspect experiment, the delayed choice and quantum eraser experiments in which the conservation laws and quantum correlates are maintained. Yet time is irrelevant in the domain of the particles. Time is only relevant in the reference frame of the observing scientists and other macro-world mortals who are as yet unable to attain the speed of light.

The modern-day social, political, and economic processes, created by humans, are founded fundamentally upon a Newtonian worldview in which time is an absolute, independent, constant-rate variable. But when we reach the edges of our macro-world map by looking at the very, very small; the very, very large; or the very, very fast; we find that time has no meaning at all in these domains. This likely has something to do with the mystic's notions of eternity.

While Prigogine was doing his work on dissipative structures, others were finding new ways to map nonlinear systems and revealing nature's use of nonlinear processes. Scientists such as Feigenbaum, Lorenz, and Mandelbrot all initially worked independently. They made discoveries in mathematics and science that helped put more meat on the bones of Prigogine's work. There are several salient facts emerging from studies in chaos and complexity theory (as areas of this new science have been called) relevant to our story of consciousness.

By looking at nonlinear behaviors in nature, scientists discovered certain repeating patterns of chaos and order, depending upon the scale of size. The closer one looks at the seemingly chaotic behavior of certain systems, new levels of order seem to emerge from the details. And beneath that level of order is another level of chaos, and so on. For example, if one looks at a large section of ragged coastline of a continent, then looks at the structure of ever smaller sections of natural waterfront, the same seemingly random patterns of roughness appear. Although at each level of observation a random pattern is evident, it is impossible to tell which level of magnification one is observing. The picture appears the same. There is a repeating pattern of sameness running from the very fine scale to the very large scale. This repetition of pattern also appears in the flow of fluids. Whether one is looking at the transition between smooth and turbulent flow in a small channel, where a seemingly random pattern of vortices and eddies appear, or the much larger scale of the flow of the atmosphere with hurricanes and tornadoes, the same patterns appear.

Revealing the relationship between the scale size at which patterns seem to replicate themselves has led to the understanding of "fractional dimensions" or "fractals," a fascinating part of chaos studies today. Discovering the fractal numbers for certain processes helps the researcher know where to look in nature to find repeating patterns of natural behaviors.

The general equation for the movement of fluids, called the Navier-Stokes equation, has always been one of the most difficult of macroscale problems to solve. Only through grand simplifications can it be solved, in very specific problems. A meteorologist, Edward Lorenz, discovered that even with greatly simplified Navier-Stokes equations, the solutions for weather patterns were dramatically changed when the starting point for computing the solution was changed even a minuscule amount. The equations mapping the flow of air masses that produce weather are so sensitive to small changes that the phenomenon was quickly named the *butterfly effect*.

The name was meant to imply that weather patterns are so sensitive that the flapping of a butterfly's wings over Beijing could cause a change in weather over New York. The amazing ability of computers to rapidly repeat a calculation thousands of times, with a slightly changed set of numbers each time, made this type of discovery possible. Before computers, the calculations for just one set of problems could consume the entire lifetime of a mathematician. But now a computer can run millions of calculations quickly to simulate the movements of energy in nature.

Perhaps the most powerful discovery in chaos theory was the discovery that a feedback process with mathematics simulates the way nature creates certain forms. Take a very simple nonlinear equation, give the variables a value, compute an answer, then take the answer as the starting point for a new calculation. Upon repeating the calculation hundreds of times, the answers, when plotted, will often trace some of the most beautiful shapes found in nature, such as a pine leaf, or a fern.

The significance of this fact is not that nature knows mathematics, but that nature uses feedback loops, in particular *positive* feedback loops, in literally thousands of its creative processes. Mathematicians use the result of a calculation and feed the result back into the next calculation to map the process of nature. But nature uses energy, a molecule, or a group of molecules to feed energy back into a new cycle of process, which results in "form." A particular flow of organic molecules in a feedback process produces a maple leaf. A slightly different molecule in a slightly different feedback process could result in a fern, or a finger, or a fig. As with the weather, a butterfly appears capable of affecting enormous change by merely batting its delicate black and yellow wings.

Perhaps the most striking idea that comes from this observation is that learning is also a feedback process. In much the same manner, we human beings observe the results of an action, then correct the input to the action in order to improve the result. We all do this every day of our lives. Likewise, the mapping of nature's nonlinear feedback processes is suggestive of learning at the most fundamental levels.

A Dyadic Model

Interconnections

19

After a few years of maintaining a comfortable distance from day-to-day operations of the Institute of Noetic Sciences and its projects, I began receiving phone calls from various board members, asking if I would consider more active participation once again. The institute's membership had swelled to more than 10,000, and with the new blood came pressure for even more diverse activity—and perhaps loss of focus. At least some people thought so. My secular viewpoint, they felt, would be healthy.

As I began to assimilate myself and my ideas into this new and larger board, I came to feel more at home. I'd routinely donated portions of my collection of space memorabilia to the institute throughout the years, which helped raise funds, and spoke about its projects to my audiences as well. A sea change seemed underway in popular thinking. By the mid-1980s, the early ideas that had begun in the margins of American society were now more widely accepted. A keen interest was forming around what 20 years ago seemed so fantastically arcane.

But I had never given up on my personal study. An area of concern had been how the belief concerning the demarcation between physical and mental events had evolved historically. This was a critical idea, as I was beginning to be quite sure that at this arbitrary boundary between mind and matter, both the classical scientific paradigm and theological thought break down. They each have something entirely different to say about how mind and body interact. It was also important for me personally because I saw how for the past 35 years I had attempted to live in both camps.

Without being fully conscious of where it would lead, I was constructing my own model of reality. We all do this in one fashion or another, tying together bits of information to make our world more coherent. My desire had been to reconcile my experience in space with my experiences with Uri and Norbu, so that I might see how they could exist within the universe we observe around us. This was a deliberate process. I wanted to know the underlying nature of the samadhi and psychic phenomena, what it was that made them at once possible, mysterious, controversial, and compelling. Like all of humanity, I had two vehicles of discovery: science and intuitive inner experience.

Years after I undertook this project I would refer to my interpretation as the *dyadic model*, as it dawned on me one day that this was the recurring concept that embodied the essence of how such phenomena tie together and operate through natural processes.[1] As I worked I tried to cull lapses into romanticism so that the fundamental structure of nature could be more honestly reexamined. The model was, in a way, a synthesis of all I had learned in the last few years, a construction of all the theory that appeared to fit together in a natural pattern. It was a summation and also a work in perpetual progress. As new information became available, such a model would have to accommodate it in one fashion or another. The model had to be humble and pliant, and never dogmatic.

I devoted my studies to those basic sciences that concern themselves with the structures of the universe, all the while looking for paradoxes and inconsistencies among existing theories. Through understanding paradoxes and anomalies, new discoveries are made. I wanted to become intimate with the velvety blackness that I'd felt so connected with on the way home from the moon. So I studied both the physical and the mystical worlds simultaneously. But I recognized that the mystical (in other words, mysteries transcending ordinary and accepted human knowledge) cannot be experienced intellectually any more than one can learn to swim on dry land. But hopefully, the experiences could be described better using tools of modern science. The preparation for Shamanistic knowledge was as arduous and time-consuming as any journey into deep space. But it was also just as rewarding.

At the time I believed it would be useful to chart the sequence of events in our evolution that resulted in the kind of thinking we experience today. I thought the sequence itself could have much to say about the way our modern beliefs were formed. This quickly turned into a long process of winnowing bits and pieces from endless sources of information. A large part of the problem lay in bringing together cohesively that which had been thought of as separate for centuries.

I discovered that this division wasn't present in intellectual thought before the Aristotelian and Platonic period in Greece. Plato conceived of the real world as the world of ideas and perfect form, and the world of our senses as only a shadowy projection of this perfection. Platonic thought was carried forward by the Gnostics to the alchemists of medieval times. It acknowledged the ability of mind to create all physical events and was the origin of the Idealist philosophy. He correctly reported in the *Parable of the Cave* that if we have limited information, we have limiting beliefs about reality.

Aristotle, Plato's student, contrarily proposed that only information derived from the normal senses enters the mind. This idea was picked up and refined to allow supernatural intervention by Thomas Aquinas, which became the prevailing philosophy of the Roman Catholic Church for centuries. As the dominant force in the West, the Thomistic structure of logic provided the rationale for flat-Earth and geocentric beliefs of the period. Ecclesiastic power provided justification for the persecution of opponents; thus, learning was controlled by Western theology for centuries.

After undergoing subtle changes and refinements, Thomism reached full fruition as a result of the dualistic philosophy of Rene Descartes early in the 17th century. Spirit and matter, Descartes concluded, were of two independent, noninteracting realms. Interaction between the two was limited to the transitory indwelling of the "soul" in humans and an occasional "miracle" of supernatural origin. Being a renowned philosopher and mathematician within the Church, his conclusion of dualism legitimized research into physical phenomena without ecclesiastic oversight.[2]

But soothsayers and witches could still be persecuted, as their abilities were demonic, and neither human nor divine. So it was in this spirit of tolerance that science and religion began to progress down separate paths. Today the policy of noninterference and peaceful coexistence is still at the core of our Western thinking, in spite of differing fundamental assumptions and mainstream scientific discoveries of this century that point to the need for integration. It is apparent that characteristics attributed only to the spirit world are likely inherent in matter itself—ephemeral, interconnected, ubiquitous, and creative. And it is equally important to point out that many known characteristics of evolved matter are the same as those that would be needed in a spirit world—existing, sensing, thinking, reasoning. Although Plato and Aristotle stand at the head of a divided road separating internal and external sensing, it was Aquinas, and then Descartes, who perpetuated and accentuated the division into two separate realms of mind and matter. But mind and matter are not separate realms; rather, two inseparable aspects of a single evolving reality. They are, essentially, dyadic.

For the past three centuries Newton has served as the bedrock of Western scientific thought. His intellectual foundation rests upon the Cartesian duality, which has allowed a materialist philosophy to arise, and with it epiphenomenalism, the doctrine that consciousness is a byproduct of the laws of physics and biology. All theologies, on the other hand, presume that consciousness—god consciousness, at least—is the preeminent, fundamental "stuff" from which the universe is structured. In the most radical of theological views, matter itself is but an illusion. Thus the peaceful coexistence between classical science and theology has had its roots for 300 years in a seemingly fundamental conflict about the nature of consciousness.

But science has finally probed deeply enough into the structure of matter to discover only two things: empty space and energy. And even the empty space is now believed to possess an energy—vacuum energy, or what is also referred to as the zero-point field. The zero-point field is defined as the field of quantum fluctuations that exists at a temperature of absolute zero and fills all space. It has also been interpreted as that field of energy that underlies and is in dynamic exchange with all matter. This is the basic, infinite, unstructured quantum potential from which existence arose. Even if the quasi-steady-state theorists are correct, and matter is continuously created in numerous galactic clusters rather than the single point of creative origin called the big bang, the field of quantum fluctuations still underlies existence of all matter. Thus, everything we know (and everything we don't) arose (or arises) from the zero-point field of energy. (Some writings use the term *nuether*.)

Beyond the range of observation available with modern instruments, nature still appears to be, for the time being, ineffable; which is to say, it is not completely describable. However, work continues unabated to discover measurable properties of the zero-point field and to observe very high-energy interactions of particles. But we've discovered that matter is interconnected nonlocally and "resonates" in same mysterious manner throughout the entire universe, and that patterns repeat themselves as though a template were being used over and over again at different fractal scale sizes. To top it all off, there seem to be basic feedback loops that suspiciously resemble the process of learning. Such abstract descriptions of nature, however, are constructs of the human mind, and have existence only within the vault of our own consciousness. In other words, they may or may not be accurate models of what nature is actually doing. After all, the mind itself is an evolving product of nature. Only by testing our maps against the territory of nature, as technology evolves to permit it, can we become more certain of reality.

To reveal the nature of mind while simultaneously probing the mind of nature requires a single, interconnected, interactive approach quite different from that of Descartes, Newton, and Einstein. Science has entered the realm where the abstract, ineffable, and ubiquitously interconnected prevail—the realm previously dominated by theology alone. But we can no longer keep the two separate in our thinking: Matter, which has been considered the real reality, is at its bottom nothing but empty space containing energy—a mental abstraction; and mind, which has been considered undependable, ethereal stuff, is our only source for discovering "reality." Together they point to nature as being but a single reality, yet one with two related aspects: physicality and mentality—or, in other words, existence and knowing.

The evolutionary process that created everything we know provided us with six sensors to receive information from the external world. One of the sensors is an internal "feeling," which is still ill-defined scientifically. The feeling sense in humans is clearly associated with intuitive processes, subconscious memories, nonlocal perception of information, and of course, internal evaluation of the state of physical well-being. But beyond this meager description there has been little detailed analysis in scientific literature. Even modern medicine, after asking where it hurts, promptly loses interest in this sense (psychiatry excepted). Fortunately, in the last few years, biology and medicine are now seriously investigating below the traditional level of chemical functioning of the body to uncover the underlying electromagnetic and quantum processes.

I came to realize that the internal feeling sense is likely the most ancient and primitive of an organism's information management processes, likely originating with nonlocal resonances at the molecular and cellular level of simple organisms. Before linguistic capabilities were well developed in early humans, we already had the capacity for this feeling sense, creativity, imagination, and problem-solving. The evidence lay in the fact that animals far less complex than ourselves respond to internal feelings. We call it instinct, but it is nevertheless an internal sensation that drives the horse to seek water. Visualization, creativity, and problem-solving are all required to create tools, cooperate in the hunt, or to organize socially—all of which are observed in the animal world, even in species far removed from the primates, our nearest evolutionary kin. This is clearly indicative of an inner life of significance in all species, well before the linguistic period of humans. So we can infer that primitive humans, and likely all animals with a brain, have an internal representation of the external world. As language was developed, words were created to represent certain internal pictures. We often think these labels are actually attached to the world around us, but complications arise by this way of thinking, as there is then a tendency to believe that they represent the absolute meaning of things. Our names for objects and ideas are only descriptions of our internal representations. We only know what the world outside the vault of consciousness looks like; how it tastes, smells, sounds, and feels. But we don't know what it actually *is*. Our sensory information only provides a limited set of clues as to what lies beyond the domain of our selves.[3]

Tens of thousands of years after developing the spoken language, humans learned the value of the written symbol. Written symbols represent the symbols of the spoken language, which in turn map the images and thoughts the mind/brain creates. Human communication, both oral and written, is then reintroduced (fed back) through the senses into the thought process to create another cycle of internal visualization and thinking. A tangled hierarchy of symbolism results from this nonlinear feedback process, which is really just a hierarchy of information. Unless care is taken to note the sources of information, it becomes a jumbled mix, an alphabet soup. Art and music are interpreted emotionally with prelinguistic functions, bypassing this jumble.

Nevertheless, the mind attempts to imbue this jumble with meaning, to rearrange it, make sense of it all. The rational, thinking portion of the brain wants it to form a logical, coherent, and consistent structure. The rational, thinking function of the brain very likely evolved specifically to help manage this tangled hierarchy of verbal symbolism. The mind makes connections and gives meaning in its effort toward rationality. But the "reasonable" structure that one brain creates is not necessarily the same as what another creates, which gives rise to the vastly different interpretations we humans give to the same events. We often find other "reasonable" structures totally unreasonable, and here you have the root of all human disagreement and conflict. The brain creates *internal* consistency and order, not *absolute* consistency and order.

Today we know that our reasoning process begins in the frontal lobe of the left hemisphere, which is a more recent accomplishment of the evolutionary process. The linguistic developments likely accelerated a need for self-reflection and critical analysis, which emerged only a few millennia later. When we use our well-honed intuition, we call into play the extensive pattern recognition and holistic functions of the right brain, somewhat independent of the messy reasoning and language processes. We err when we attach absolute significance to the symbolic representations presented to the internal screen of our awareness, either rational or intuitive. We also err, as did Descartes, when we assume that our internal states evolved differently than those of our animal cousins.

Our human descriptions of external reality are a culturally defined consensus on the meaning of the images and symbols central to that culture. Such consensus is arrived at through the exchange of informational symbols. Therefore, "knowing" is just a meaning (another internal label) we attach to information when there is a feeling of certainty about the assigned meaning of that information. But additional information can easily invalidate our knowing. This is why our better judgment prompts us to seek the second opinion, or to seek independent validation.

20

Having studied the physical and the mystical simultaneously, I saw parallels emerge from the data, similar patterns of structure between religions and this new brand of physics. They are subtle, but nevertheless there.

The origin of all religion is rooted in the mystical experience. From the earliest Shamans and tribal medicine men, humans of all cultures have discovered that under certain conditions one can seemingly perceive information from beyond the immediate environment (nonlocally).[1]

With the help of ritual to allow the mind to transcend the mundane, the Shaman can see with the eyes of an eagle and discover the approach of an enemy. He or she can commune with buffalo and assure the success of a hunt, and resonate with the flora to discover healing herbs. In my travels through the years I've had the privilege on many occasions to visit with Native American medicine men, the Kahuna of Hawaii, Shamans of the South American tribes, and Voodoo priests of Haiti.

What I've learned firsthand is that the similarity in their understanding of these extended human capabilities is remarkable, though not coincidental. The differences are in the cultural metaphor and the details of the rituals they use—rituals that spring from the environmental setting of the people. That is to say, the differences in attaining altered states of consciousness are superficial, but the cultural mindset, interpretations, and ritual that spring from their traditions may be vastly different.

When I met with the Kahuna of the Pacific islands a few years ago, I noted how they greeted me with easy smiles. Their manner of movement and thought seemed to reflect a harmonious kinship with the land and sea of their lush tropical islands. One evening I spoke with a few of the older natives who told me of what they call the "psycho-navigation" techniques of their ancestors. Generations ago they crossed vast ocean distances in frail boats, and unerringly found small islands using their knowledge of wind, sky, and sea—but mostly by using their finely honed intuitive skills. They spoke most easily of kind and loving spiritual connections to all life as fundamental to their existence. This is reflected in their graceful, sensual dances, and the feasting of their ceremonies and celebrations.

Similarly, the Native American Shamans of the Southwest tribes emphasize the interdependent relationships of nature, with rituals of purification for the mind and body before calls upon the spirit world for insight, guidance, and sustenance. But their land was more harsh and foreboding. Likewise, their demeanor expressed a more stoic and determined aspect toward life, the spirit world, and the rituals that help bring them together. The dances and rituals of supplication seem to have a greater intensity and urgency in this harsher land. Among the Voodoo priests of Haiti, the frequent mood changes in blood-letting rituals and animal sacrifices speak of a more foreboding side of the psyche. The malevolent spirits are to be appeased and enlisted in the human struggle for survival, and then safe passage for the soul through the terrors in the afterlife. Dances that express emotional abandonment, and the inducement of trance states, create an electric aura of mystery and dark otherworldliness among the ritual participants.

But they all have in common the belief that the visions perceived in trance and dreams are exactly what they appear to be—assistance, obtained through either benevolence or supplication, from a spirit world. Great effort is made through sacrifice, ritual, and offerings to maintain that cooperative relationship with natural spirits, usually for the same reasons and with the same sense of commitment that prompted Abraham to offer his son Isaac as a sacrifice to Jehovah, as recorded in the 22nd chapter of Genesis. The fundamental notion is to keep the spirits happy at all costs, as they are more powerful than we. Visions of this interior world are just as real and important as physical events on the exterior. And one leads into the other. Religion may be described as the exoteric cultural interpretation of the esoteric mystical experience, which arises from within the pre-linguistic functions of the brain.

The interpretation, or meaning, of the mystical experience began to change noticeably from uncritical acceptance to examination, analysis, and understanding, in four different cultures in the sixth century BC. Philosophy arose.

Under the influence of Lao Tse in China, Gautama Buddha in India, Zoroaster in Persia, and the early Greek scholars, critical analysis took hold. Within decades of one another, each began to question independently the nature of the inner experience. And like the parable of the four blind men examining the tail, leg, ear, and trunk of an elephant, each culture emphasized a different aspect of the whole, though there is remarkable agreement as to the virtues required for the well-lived life.

The Taoist approach observes the interconnected fabric of the whole of existence, and emphasizes placing oneself in harmony with the movements in nature. The followers of Buddha learn to control the inner experience through disciplines of the mind and detachment from physical desire. Zoroaster learned to harness the power of intentionality to influence the course of natural events, producing followers of great influence. From Zoroastrian Persia came the Magi to Jesus's birth, presumably sorcerers of great capability, from which came our term *magician*. The Greek scholars concentrated on the rational thinking capabilities of the mind, though it is believed that Socrates was also an adept in the Zoroastrian school in Persia before reaching eminence on his own in Athens.

These four great schools of thought, when taken together, complement each other and contribute to a broader understanding of the capabilities of the mind and consciousness. All are required to help complete the picture. This evolutionary sequence clearly suggests that our emotional responses to any information certainly preceded development of the intellect. And emotion may be considered as an energetic response to an internal feeling.

In the Middle East another extraordinary development had already taken place. Here the Hebrews had consolidated the early Shamanistic reliance upon multiple natural spirits and deities into a single deity with the forbidden proper name *YHWH*. Of course, the Semitic culture that produced the Hebrew tradition later spawned Christianity and Islam, both of which center around the idea of a single omniscient, omnipotent, creating, anthropic god.

As I studied the similarities and the complementary features of the world's mystical traditions, it occurred to me that the mystical encounter by Abram of Ur, later Father Abraham, with the one omnipotent source of creation, might have been an esoteric insight presaging the idea of conscious energy as the single source of existence. Steeped in the tribal Shamanistic tradition of prior centuries, however, his insight was interpreted, given literal meaning, and explained as external, otherworldly, and anthropic. Today we might interpret it as an internal vision.

A radically different explanation for the one-God concept is found in the writings of Zecharia Sitchin, a Sumerian scholar. Though largely ignored by other Middle Eastern scholars, Sitchin's translations suggest that the account in Genesis of creation is an abbreviated version of tablets unearthed in Sumer and Chaldea. These tablets record a more complete story, which also includes planetary configurations not known until this century, the details of which were only confirmed by spacecraft flybys of the planets during the last 25 years. If valid, it's an astounding bit of news. Science requires that a theory make successful predictions, and the Sumerians may have done just that long before we had a means to confirm them. Unfortunately, Sitchin's translation suggests extraterrestrial presence in the prehistoric period as the source of planetary knowledge and the one-God concept—a totally unthinkable idea a few decades past, but not necessarily so today if the translations are correct. However it began, the Hebrew tradition did bring to the world a one-God concept, which is fundamental to three of the world's great contemporary religious traditions.

A common thread that runs through Taoist, Buddhist, Zoroastrian, and Greek thought from the sixth century BC is the ability of the mind to be trained to enhance and enlarge the inner experience of the student. While the Greeks of the Aristotelian school emphasized rational, sensory input as the information source for deep thought, the other schools, including the Greek Gnostics, emphasized the mystical, nonlocal experience as the source for insight. In the parlance of the 20th century, Aristotelians alone placed emphasis on sensory input from the five normal senses, and left brain, linear thinking. The other schools emphasized training the mind to better perceive and utilize the intuitive (mystical) input of information—a right brain (or prelinguistic brain) function. And Orthodox Judaism, Christianity, and Islam emphasize unswerving belief in an anthropic-style Creator.

I found it ironic as I studied the historical evolution of the thought process that in the scientific world we often make the same error the earliest Shaman did. The Shaman perceived an internal vision, took its meaning to be obvious and literal, and assumed that the internal visions and the external world were one and the same. Likewise, the modern-day scientist looks at the external physical world, creates an internal picture, usually with mathematics, and assumes after some experimental validation that mathematics is actually embedded in external reality.

In both cases there is a subtle but important difference between recognizing that a meaning has been *given* to information derived from nature, and believing we have discovered some external truth *inherent* in nature. Only by critically examining the processes of mind and how it uses information to form beliefs, maps, and models can the distinction in either case become clear. Even today many devotees of the mystery schools suggest that mind, or thinking, is a block to attaining enlightenment or union with the godhead. One must put aside thought and rely only on the trained intuitive experience to perceive the ultimate reality.

But this notion is only partially correct, for we need thought to help decipher the tangled hierarchy of information and apply meaning to the insights. Modern studies suggest that it is indeed suppression of the 12- to 24-cycles-per-second beta waves that permit intuitive insights, visual imagery, and mystical visions to occur most easily—all of which originate as right-brain functions. However, learning to create balanced whole-brain functioning and coherently utilizing the entire brain mass is likely the most effective technique for learning. Rather than neglecting the thinking functions of the brain in order to reach the intuitive, or neglecting the intuitive to enhance the thought, both functions enhance the whole when optimally and coherently trained.

Einstein once said that science without religion is blind, and religion without science is lame. Here he hit the mark. Science relies heavily upon left-brain, linear functions, while religious experience relies more upon visions and intuition. This suggests that it is the complementary and coherent utilization of both of these evolved faculties that must be understood, enhanced, and practiced. Emphasis cannot be placed on one to the exclusion of the other. Again, a fine example of a dyad.

It's not without reason that the mystical traditions have taught meditation, silence, and self-reflection. This calm, quiet state of awareness produces insights not available when the brain is rushing at a high energy state of 24 cycles per second or faster to manage our daily routine. Whenever successful meditative disciplines are employed in the pursuit of insight, they result in reduced stress, an elevated sense of calm, buoyed spirits, a greater awareness of synchronous events, and a heightened awareness of nonlocal information. It also appears that in some cases, after prolonged practice, the individual can become psychokinetically active. Modern evidence suggests that all children can be taught to enhance these faculties through simple games and exercises.

Early in my studies, questions continuously arose concerning the relative capabilities of earlier and modern human beings. What is it that has evolved and changed within our thought process? The evidence is rather unequivocal that a gross measure of intelligence (the brain mass) has remained relatively constant since Cro-Magnon times. But the internal capability—the "firmware" and "software" that produces thoughts, capabilities, beliefs, and knowledge—has developed dramatically during recent millennia.

The brain is the only organ of the body to have produced capability well in advance of the need—a remarkable occurrence. We are still evolving into our brain, or better said, we are still learning how to optimally program this magnificent instrument. Evidence from those who survive into adulthood with the affliction of hydrocephalus, a birth defect in which only a small amount of brain tissue is present, demonstrates that normal human functioning can be accomplished with only a tiny fraction of the brain mass available. It would seem that most of us use our brain mass quite ineffectively—a Cray supercomputer even to balance the checkbook. But the hydrocephalics that survive must be using their brain mass at near-maximum capacity by comparison.

The amazing capability of the human brain/mind to perceive the most minute detail, and then the broad scope of the heavens—to look inward, then outward and observe—and to contemplate meaning, is perhaps nature's most astounding biological achievement. But this hasn't always been so. Only within the last few millennia has the self-reflective capability evolved and reached a stage similar to the one we experience today. How the brain perceives information and imbues it with meaning is the critical issue. We err when we accept uncritically and literally the experiences and understanding of yesteryear as equally valid and applicable in today's world. In an evolutionary universe this simply cannot be true. Here change is fundamental, even though on galactic scales the time frame is billions of years, decreasing to centuries with the presence of simple life, and decades with anthropic beings. This changing time scale alone suggests the importance of complexity, brain function, language, and self-reflective reasoning.

21

As I studied the beliefs of mystics, and psychic phenomena, it became absolutely clear to me that first-person experience, the subjective, with all its potential for misinterpretation, was just as important to understanding reality as the third-person observations of science. Nature has provided a broader range of mental capabilities than can be captured within the norms of Western cultural tradition. As Willis Harman had suggested to me in our ongoing dialogue, we needed a better way to validate these first-person insights.

I suspected modern research could be useful in revealing methods whereby years of ascetic discipline required by traditional "mystery" schools could be abridged and made available to more people. In looking for a kind of shortcut, there were also traps. At the time, the use of recreational hallucinogens was popular among young people. Though natural hallucinogens have been used for millennia by Shamans of various cultures, I knew this was such a dead end. The risks were far too great, and the benefits dubious unless conducted under rigorous control and for purposes more serious than just "getting high."

Perhaps the most thorough and detailed mapping of inner experience comes from the Buddhist and Hindu mystics. The Tibetan Buddhist monks in particular have approached the subject with scholarly intent and precision for centuries.[1] The most exalted state of awareness is described in the mystical literature as the *nirvikalpa samadhi,* a name derived from the ancient Sanskrit. This is a state of awareness in which there is only Self;[2] that is to say, only the transcendent observing entity. There are no thoughts or objects in mind. Indeed, Self expands and merges into the entire field of mind so that pure awareness is all that appears to exist. The state is accompanied by an ecstasy that seems to permeate every cell of one's body, and results in a feeling of certainty about the eternal nature of Self. Beyond this simple description, the state is ineffable, which is to say the description falls short, and doesn't assist others in attaining it (though it does help one recognize the experience when and if it occurs). The state must be experienced to capture its complete essence, however.

In Christian literature the phrase "the peace that passes all understanding" is often used to imply the ineffable character of the experience. The theological meaning often given to this samadhi state is that of "union with the godhead," or, to use Tillich's phrase, "union with the ground of our being." I would suggest, however, that the meaning assigned is not inherent in the experience, but rather is the result of attempting to describe the experience in accordance with one's theological beliefs. We do not "see" God in such an experience. Nor do we experience union with God—unless we are already predisposed by prior belief or training to expect that this is what the experience means.

My own efforts to experiment with this state throughout a number of years led to the conclusion that a lifetime of ascetic discipline was not the only path by which one could naturally experience these altered states of awareness. By combining modest changes in diet, routine meditation, and breathing exercises, along with a certain level of detachment from the pace of daily modern life, I found I could experience successively greater involvement of my entire brain/body with the samadhi experience—and achieve increased awareness of the effect of participation of the entire body in the experience. The mind is devoid of thought and images, yet acutely aware and alert. Each cell of the body contributes intense sensations of pleasure and well-being, the sum total providing an enveloping aura of bliss or ecstasy. Although the presence of the Self as the observer is implied, there is actually no notice of Self during the experience. Awareness is so flooded with the sensations of joy, universal connectedness, security, and well-being that Self goes unnoticed. It dissolves into the experience.

Individuals who report that they have had mystical encounters with divine beings in certain altered states of awareness are convinced of the "reality" of their encounter. Indeed the vision is real enough, but those who continue the journey and then reach beyond the symbolism of those perceptions, may find themselves catapulted into the undifferentiated awareness of nirvikalpa-level samadhi. Only then are they able to recognize that the prior encounter was from culturally derived images arising from within the subconscious, permitting conscious awareness to interpret a visual image as a mystical encounter with another being.

The atheist experiences samadhi as well, but prefers not to describe it in theological terms; rather, in terms of the sensations and images the experience evokes. In this sense, *ineffable* is similar in meaning (with regard to mystical experience) to *uncertainty* in quantum physics. For both mystic and scientist, the deeper reality is difficult to describe in the language of the everyday macroscale models with which one attempts to understand and communicate reality. The mystic revels in the mysteries of the ineffable, while the scientist chafes at the lack of specificity. Though the map cannot be the territory, exploring new territory helps us construct a better map and reduce the seemingly ineffable nature of the experience.

When attempting to understand the nature of consciousness we must keep in mind that we do so with a highly evolved anthropic organism: our mind/brain/body. But certain aspects of consciousness and conscious awareness are not unique to human organisms. Other organisms, and perhaps all matter, are in some sense conscious and aware as well. One of the more significant flaws in traditional Western thought has been to view ourselves as qualitatively different from other species in this regard. This has, in turn, impeded our understanding of both nature and consciousness. However we experience our consciousness, the evidence suggests that it has evolved from more primitive beginnings, and didn't spring full-blown to the states we experience today. By studying other earthly lifeforms we can gather clues to our more primitive nature, and to the processes involved in evolution of the anthropic consciousness we experience.

In the Idealist belief system, where consciousness is the only basis for reality, the nirvikalpa samadhi is viewed as the beginning of creation and the end of human striving. It is both first and last. Although I believe the Idealist structure is fatally flawed for a number of reasons, it offers a superb method to examine the psychological structure of different states of our conscious processes. While in nirvikalpa samadhi, the creation of an object (a thought) would then be seen not as Self, but as something separate from Self.

This is the first dualism: The distinction between the created object and the Self that created it. The Self now becomes the self-aware observer, or "self" of everyday experience. Creating this dualism by observing things as separate from Self causes one to leave the samadhi state of consciousness and arrive at the state where "things" exist. This is termed the *existential state*. Perceiving things as separate from Self can occur only at the existential level or below. The subject/object dualism separates and distinguishes the existential state from the samadhi states, and "creates" the world of illusion where things are separate and distinct from each other. In this fashion, God, or the gods, create the world. In the Idealist interpretation of this experience, the world is but a thought in the mind of God.

A major island lies in the gulf between the level of nirvikalpa samadhi and the existential level. This is a level where one can observe things as separate from Self, yet recognize that they are all connected to each other and to Self, that separation is but an illusion. This point is called the *savikalpa samadhi*, and it too is accompanied by the experience of ecstasy and eternity. It is the state I spontaneously experienced in space as we returned from the moon, though I didn't know then what it was called, or even that it had a name.

The ancient mystics who first explored these psychological states defined samadhi as the god states, because they came to believe that all creation flows from the arising thought. The other states result from creating additional dualisms (illusions) that separate one from the purest state of awareness. All humans have likely experienced samadhi, if only momentarily, in life. We occasionally catch glimpses of those ecstatic states in which only unity prevails within the Self. If they were more rare, it would be a pity.

Experience has taught me that much of human motivation is a subconscious drive to reexperience the ecstasy of the samadhi. I suggest that samadhi is the prototype for the pleasure side of the fundamental pain/ pleasure response that drives all animal behavior—human or otherwise. But the memory usually isn't in our conscious awareness, only in the subconscious as an archetypal memory. Perhaps the initial experience for each of us lies in the womb, where we are only matter with undifferentiated awareness, not yet self-aware, but experiencing and storing information in the subconscious from within a warm and ecstatic watery world. It may even have more fundamental origins.

The ancients discovered that between the existential state and the samadhi states are the levels of consciousness where the psychic event takes place. It's in this psychic realm of mind that information is perceived nonlocally and direct action is manifested in the physical world— that is, with psychokinetic effect. The psychic states that exist between the existential and the samadhi states are less acknowledged, particularly in Western civilizations. Whether they are less experienced or merely less acknowledged is debatable.

For strong psychic events to occur, such as what I witnessed with Uri, Norbu, Shamans, and the numerous children we collectively investigated, there seems to be the need for certain well-developed neurological connections in the brain. Modern neurological studies suggest we all have those connections in childhood. In fact, nature provides a wider range of brain capabilities than any particular individual is likely to use. But in the absence of exercising the ability, they wither and are lost. We either use them or lose them. However, to a certain measure we each create our physical realities day in and day out as we manage our internal states. Thus, it seems, many of the more subtle manifestations of the psychokinetic capabilities are ceaselessly functioning at subconscious levels for each of us. When one intends to lift a finger, one's finger instantaneously responds. Medical science models this as due only to the connections of nerves, muscles, and tissue connecting the finger and the brain. However, a strong case can be made that deeper quantum functions come into play and that the classic medical model is incomplete. Intention operating within the body, such as lifting a finger, is not significantly different from an intention to manifest an effect outside the body, such as Norbu's healing or Uri's abilities. One is directed toward a seemingly local object, the other toward a nonlocal object. From the point of view of the brain, however, both are nonlocal. It is only our entrenched belief that classic models adequately explain physical (or spiritual) functioning that prevents discovering a deeper reality, such as my mother rejecting Norbu's healing, then slowly recovering, regaining her sight over the course of several years.

The state of consciousness just below the existential is termed the *ego state*. It is within (or below[3]) this state that most of us find ourselves when awake. The ego state is reached from the existential state (in the Idealist structure) by creating another dualism. This second dualism is characterized by losing or forgetting the sense of the eternal, and is accompanied by the loss of ecstasy. In the Christian religious literature, this dualism corresponds to the fall from grace, humankind's separation from God. A beautiful allegory for this is the story of the Fall as recorded in Genesis, when the first two humans ate from the tree of knowledge in the Garden of Eden.

The psychological interpretation of this process is that the state of subject/object existential consciousness is further divided upon recognition that one inevitably approaches physical death. The cognizance of death, together with all the fears the recognition evokes, marks the difference between the existential and ego states. Whereas the existential state observes the separateness of things, it *remembers* the ecstasy, the connectedness, and the eternal nature of the Self. It is also capable of assuming an amused, detached perspective of the world. The ego state, having lost ecstasy, the sense of connectedness and eternity, is a state of perpetual fear and insecurity. It is here (or in the lower states) that we live most of our lives with the fervent hope that this is not all that life has to offer.

The ancient mystical thinkers created these remarkable maps of the states of consciousness without the benefit of precise knowledge of physical evolution. They had no way of verifying whether their maps included all of the territory, or whether human consciousness could be evolving in some way. I learned, however, that there is a physical interpretation of these states of consciousness. It wasn't until I began to add physical evolution to the mystical map of consciousness that my own picture began to change in subtle ways.

The Neanderthals left the first record of ritual burials, and because of this there is reason to believe that the evolutionary development of ego state consciousness may first have occurred for humans during this epoch. However, other species mourn their dead as well; elephants, for example, make their final pilgrimage to particular locations to die. But mourning the dead and ritual burial are not necessarily tantamount to the personal cognition that "I" will surely die, though it should eventually lead to that cognition. If we take the development of a child as a personal reenactment of the evolution of our species, a child first has undifferentiated awareness, then becomes self-aware at several months of age, and may then experience the pain of the loss of a loved one. But only later does the child relate this experience to its own life and begin to fear its own mortality. At this point, self-reflection has already begun—the examination of the content of awareness.

By this way of thinking it seems reasonable to suspect that evolution of the ego with individual recognition of inevitable death likely came about later rather than earlier for humans. Certainly it evolved after self-awareness, for it requires a certain measure of self-reflection, which is a later development. Self-awareness and the ego state of awareness are not one and the same. The critical issue that arises here concerns the separate existence of the Self. Is there, as the ancient mystics taught (and most still believe today), a "hidden observer" within each of us that is nonphysical—a "ghost" in the physical machine? We seem to experience this Self, this inner observer we label "I." But is it real? Can it exist apart from our experience of physical reality?

When Descartes stated, "I think, therefore I am," he set the tone for modern Western perspective on the issue. He also concluded that body and spirit were of two different realms—that the soul dwells within the machine, but only temporarily. The mysterious aspect of our awareness is that there is no locus in the brain for this sense of I. Science has found no evidence of an independent observer directing the process, nor a way for it to interact with physical reality, were it to exist. Neither have the Tibetan Buddhist scholars (almost alone among mystical scholars) proposed that there is a separate Self. Virtually all attempts to find or to model the seat of existence result in an infinite regress; that is, a smaller and smaller homunculus (little being) within each level, which serves as the seat for outer levels of existence.

The act of recognizing that there are things that exist separate from Self allows us also to think "I am" or "I exist" (when in the ego state or below). Before this cognition, awareness is undifferentiated, without a distinction between Self and other. The organism is merely aware matter—it isn't yet self-aware matter. There are simple responses to pain/pleasure stimuli and the storing of information while still in the womb, but not a sense of "I am."

Both the development of a human child toward self reflection and the evolutionary history of the species seem to support this idea. The sense of *I* is the result of the aware matter's need to coherently organize sensory information. In doing so, it finds that certain perceived information originates externally and is not a part of Self. In this process the external sensory information tends to obscure the more subtle internal sensing and nonlocal resonances that dominate in the samadhi states, and provide the sense of connectedness and eternity. The sense of Self only emerges after it begins to experience the environment with the five physical senses and discovers certain things that do not appear connected to it. That is to say, we existed physically before any personal recognition or contemplation about our existence, both as individuals and as an evolving species.

The ancients provided a rather accurate map of the various internal states though they had no data that suggested there were parallels in human mental development with biological evolution. As physical organisms evolved greater complexity they also required more complex ways to manage the information available in the environment. The scientist would say that life on earth became more "intelligent" as it moved up the evolutionary chain toward *homo sapiens*. This is a third-person point of view. From a first-person point of view, we can say the progressive dualisms that separate the mystic's states of consciousness have counterparts in the physical evolution of consciousness itself. Both matter and consciousness become more complex as they evolve. The mystic's map of consciousness states is concerned with "knowing." The ancients, however, had no way of recognizing, as we do today, that existence evolved simultaneously to produce our ability to know.

The Idealist views consciousness as progressing downward from the divine experience to the mundane. In between lay a series of dualisms (illusions). If we interpret this progress as moving from the most simple toward the more complex, it mirrors the progression of physical evolution. That implies, of course, that the experience of awareness (before self-awareness) is actually the most fundamental experience of consciousness. It is inherent, at least, in all living organisms—even the simplest. Should we discover that Self can exist apart from physicality, as most esoteric traditions suggest, the mechanism should be the same among all self-aware creatures.

As a boy I was impressed with the connection my father maintained with the animals on the ranch. His bovines displayed personality traits he was infinitely acquainted with. They clearly seemed to display rudimentary self-awareness. He resonated with the animals in ways that I later came to recognize in Shamans I studied. Though he never thought of himself in such terms, he clearly found his spirituality in the processes of nature. He could intuitively recognize if one in his herd was in trouble. In the middle of the night he might awaken and go searching for a cow that was having difficulty delivering a calf. And invariably she was. He would unerringly find her in the middle of the night even though she had hidden herself in the brush for safety. These nonlocal resonances that are the most basic of nature's information management schemes work not only within a species, as biologist Rupert Sheldrake has proposed, but between species as well.

The most modern studies with primates suggest that these animals have a concept of self in that they are capable of signaling their desires and their internal states to researchers in very subjective terms. Moreover, they have long-term memory. What we don't know is how aware they are of their own awareness. We can be reasonably certain that self-reflective awareness, which distinguishes the human capability, is a highly evolved function that isn't likely present in lifeforms with brains that cannot accommodate language. Most animals that have a brain and utilize multisensory information likely have at least a rudimentary concept of Self, as evidenced by survival-oriented behavior. Basic undifferentiated awareness likely exists in more simple organisms, and may extend all the way down the chain of organizational complexity—even to inanimate structure. At least I believe it quite possibly does, and this is the thesis of the dyadic model I was gradually piecing together.

I recall years ago when my younger brother Jay and I visited our parents on the ranch with our young families. With Jay's infant son Michael we accompanied our father on his morning chore in the pastures to feed the herd of cattle. I held the child while Jay and Dad worked in the field.

My father was the kind of rancher who was close to the land he tended and the cattle he raised—a rancher from another time in the American West. As we drove out into the pasture together he told Jay and me that he was amused by one Hereford (a breed of cattle) in particular. Such conversation surprised neither Jay nor myself, as it was a part of our upbringing. Land, cattle, weather, the progress of the seasons—this was the conversation around the breakfast and dinner tables. In any event, this favorite of my fathers was the lead cow in the herd, and I recognized her at once as the cattle filed toward us in the hazy morning light, her healthy girth easing through the ripe green grass, 50 or so other cows following her lead. She immediately came around to smell each newcomer at close range, her soft, leathery muzzle but inches from our outstretched hands. Her inordinate curiosity was the source of my father's amusement and fondness.

Normally, he told us, she would have exerted her prerogative as head of this society of gentle creatures, and been the first to eat. But upon encountering me as I held the child, sniffing my nephew with particular care, she was clearly puzzled. The infant and I, she seemed to sense, did not belong together. Without eating, she proceeded back toward my brother and father, smelling each in turn once again. Then she seemed satisfied. Having presumably resolved the dilemma as to the relationship of the child, Lenore muscled her way into the feed trough and began to eat with the others. She was, in my opinion, displaying leadership, a surprising intelligence, curiosity, problem-solving, and certainly a subject/object awareness.

In spite of our traditional beliefs that the consciousness of animals is fundamentally different from our own, those who have worked with animals recognize that at the most basic levels there are similar processes at work. A quote from Sanskrit lore suggests that even the ancients felt the need to resolve these enigmas: "God sleeps in the minerals, awakens in plants, walks in the animals, and thinks in man."

The more organized aspect of conscious functioning clearly requires a highly complex brain produced by the evolutionary process. This evolutionary aspect can be labeled as mentality. The more evolved the brain, the more complex the functions it can accommodate and perform. It is mentality that most closely conforms to the materialist notion contained in the doctrine of epiphenomenalism.

The mentality for self-reflective awareness seems to require the very complex physical structure found in the human brain. But it would also appear that simple organisms such as amoebae, which possess no brain at all, may still have awareness. Clearly they receive information from the environment. But do they perceive information? Does a sunflower perceive sunlight, or does it merely maximize that energy by continuing to face the sun through a mechanistic feedback loop?

Here lies a critical detail: the difference between perceiving information, and receiving information without awareness. I believe undifferentiated awareness exists at simple levels of organizational complexity. The true primordial state of awareness is undifferentiated before the idea of "I/thou" has arisen. We cannot, however, discover awareness directly, as awareness is a subjective attribute unique to the particular organism. The approach to awareness in objects requires inference from other observations. For example, some simple organisms that are mobile move neither randomly nor deterministically. They seem to move with intent—toward food and away from danger, toward a mating opportunity and away from rivals. But not always.

It's much easier to deduce intent by observing organisms in nature than it is to deduce awareness. Intent is suggested when movement is neither random nor deterministic, but rather appears to be with purpose. However, from chaos and complexity theory we know that complex processes in nature are only successfully mapped with nonlinear equations and feedback loops that are suggestive of learning. And learning requires a feedback loop involving awareness and intention. If there is intention in the behaviors that an organism exhibits, then there is likely awareness as well. Learning requires both. Awareness is the perception of energy (or patterns of energy: information), and intention is the volitional propagation of energy.

In my evolving model, I saw that awareness and intention form a dyadic coupling—like two faces of the same coin, they are always found together, and in the presence of a feedback loop they lead to learning. Even in the first chapter of Genesis, the ancients recognized this fundamental principle: After each creative act, "God saw that it was good." The process is quite simply one of intention, awareness, and evaluation (meaning). But if simple organisms learn, as do more evolved ones, then nature itself is a learning process.

In many religious traditions (including Christianity in its early years), subjective experience is believed to be carried forward by the reincarnation of souls into successive life experiences. The concept of transmigration of souls from lower to higher life forms may be considered both as means of conveying experience, and as a mystical metaphor for the evolution of conscious awareness.

In some cultural traditions the soul of the individual is sent back to inhabit lesser lifeforms to learn again, if behavior in a particular life is in some way unacceptable. But these concepts rest firmly on the belief that souls exist independent of the physical form they take, and that there is external judgment and decision-making deciding the fate of the soul—assumptions that seem nearly impossible to validate scientifically. However, if we think of life experiences as mere information, other ways of addressing these issues seem to open up.

What I have left out of this description of learning is memory, the storage of information. For learning to take place, memory is required. With science we came to recognize that nature stores information in a number of ways, such as the DNA code. Atomic matter itself, being organized energy, also carries information. The laws of physics and chemistry are but man-made maps of nature's processes, yet we don't know how they arose. In some way nature repeats these same processes again and again throughout the universe as though a template created them all and the information retained is to be continuously available. Science has not expressed the *mechanisms* by which nature is lawful, but has just codified the rules it appears to obey. It's quite likely that there are modes of information storage in nature yet to be discovered.

Science hasn't yet addressed how individual human subjective experience is preserved, if at all. The ancient mystics created the concept of *akashic records* to account for the retention of information about human experiences and choices. But they developed this concept to instill fear of retribution among the populations for a misspent life, not necessarily to reveal the structure of nature.

Biologist Rupert Sheldrake refers to certain modes of information storage as *morphogenetic fields,* and the mechanism to access them as *morphic resonance.* I would suggest that morphic resonance and non-local correlation of particle attributes are the same process, but operating at different levels of complexity. The feedback loops, nonlocal resonance, and their possible application to learning, flies in the face of evolutionary biology, which has held for more than a century that all evolution is random mutation and natural adaptation. In other words, the Darwinian model states that nature mutates by random processes. Those organisms that happen to best fit the environment thrive, while the others die. But nonlocal resonance is now a demonstrated physical principle, and undoubtedly plays a fundamental role in nature's information processes.

The evidence that learning processes and intentional action is present in nature and communicated by nonlocal resonance is far more prevalent than that of mere random or deterministic processes. That is, nature's ways appear to be purposeful. In the last few years the idea of dissipative structures and chaos theory has forced us to acknowledge that scientific preoccupation with simplicity and linear modeling have swept under the rug the actual complexity of nature. Likewise, the need to understand human consciousness has demonstrated to us that nature is more intimately connected, conscious, and purposeful than was previously assumed. Fractal notions within chaos theory suggest that the awareness and intention of evolved organisms is likely a repetition of awareness and intention at lesser scales of organizational complexity as well, with nonlocal resonance playing a role in how information propagates between levels of scale size.

I am not discussing "learning" in the same sense that computers are said to learn. Intention requires the concept of volition, the ability to choose whimsically, even irrationally. Computers are still computational devices that operate according to algorithms (computational schemes), though the programmer may not know where the computation will lead. Ultimately they are deterministic. I adhere to Roger Penrose's argument that "knowing" is more than computation, though for different reasons. Although inanimate matter also exhibits nonlocality, and likely even rudimentary awareness, whenever we eventually understand the phenomenon, it's virtually certain that the nonlocal characteristics of a human organism will be qualitatively different than the nonlocal resonance and awareness of the mere computer chip.

Throughout the philosophical literature and the scholarly mystical[1] literature are references to the Great Chain of Being and the Perennial Wisdom. These concepts represent enduring beliefs about existence and about the values that guide humanity. They are distilled from centuries of human experience and philosophic quest. By discussing the problems of quantum physics, I initially began at one end of the Great Chain of Being, the end where energy manifests into inanimate, though not necessarily mechanistic, physical reality. By discussing the samadhi states I moved to the other end of the Great Chain of Being, toward the perception of the god state. But in a dyadic model, these two ends are joined.

I perceive the Great Chain of Being as an interconnected, self-organizing, evolutionary system. Its structure mirrors that of the universe. It's a chain connected to itself, not a hierarchical structure beginning with matter at one end and leading to God at the other. The sciences from physics to neurobiology are doing a splendid job of ferreting out the secrets of the physical aspects of matter, but will doubtless do a much better job when the role of consciousness is brought into the equation as the unifying concept.

The Great Chain of Being is not synonymous with "the spectrum of consciousness," except in the most radical interpretation of Idealism, such that all physicality is but illusion. The spectrum concept emphasizes both the ability of the evolved brain/mind to perceive and reflect upon all states of matter and mind in addition to implying the existence of consciousness in matter other than the human brain. The Great Chain of Being emphasizes the connectedness of all existence from inert matter to the godhead. I am taking these ideas one considerable step further.

The idea of consciousness as dwelling in matter and perceiving matter is best expressed through the notion of the spectrum of consciousness. Perennial Wisdom is meant to suggest the enduring ideas and knowledge

From Outer Space to Inner Space

prevalent in all ages and cultures, knowledge derived not only from the rational thinking process, but primarily from deep within the mysteries of the intuitive. Concepts of the Perennial Wisdom emerge, though somewhat differently, when the Great Chain of Being is thought of as "closed," self-organizing and evolutionary, and knowledge thought of as information perceived with awareness and given meaning within consciousness by its relationship to other information.

Though I've used the term *consciousness* in its normal English language usage up to this point—that is, self-reflection—it is necessary to be more definitive and precise with diction. Consciousness has evolved into broader usage in recent decades as a result of the influence of Eastern thinking in the West. Today it implies the broad scope of mental phenomena—awareness, intentionality, and problem-solving ability, as well as its strict meaning of conscious awareness, or being aware of being aware. Therefore the term *mentality* is necessary to include those functions of the brain such as problem-solving, self-awareness and self-reflective awareness, which are dependent upon the evolutionary complexity of matter. Awareness, or undifferentiated awareness, implies an attribute of nature, which I believe is not reducible to brain function, but more likely associated with all matter, its nonlocal properties in particular. *Intention*, as well, is an irreducible concept.

If indeed the origin of the universe was neither random accident nor deterministic, it must have been intentional. The irreducible aspect of intentionality is demanded by the fact that we experience our intentionality, and that direct psychokinetic events do exist; they are most likely latent in all human beings. But if it is latent in us today, then it must have been so in pre-humans as well, and before that in morphic nature, and so on back to the beginning of the universe. Intention cannot directly influence matter unless the potential for intention resides as a basic attribute of matter. Inanimate matter cannot be directly influenced by its complex product, the brain, without intentionality. This concept caused the creation of the dogma of epiphenomenalism in materialist philosophy, and the belief in supernatural occurrences within cultural religions. If we validate psychokinesis and find evidence that it is a general human capability, as many researchers now conclude is the case, then we must also conclude that its origin is within the fundamental structure of the universe.

Note, however, that it is necessary to use the terms *awareness* or *undifferentiated awareness* and *intention* much as the quantum physicists were forced to use *particle* and *wave* to describe phenomena they could neither see nor measure. In the macroscale world, we experience our learning selves as needing both awareness (cognitive ability) and intentionality (volition) in order to function as we do. Whatever the subjective counterparts of these labels at levels of natural organization less complex than a brain, we can hardly imagine. Besides, we don't have a language to describe them. Presumably we could use apples and pitchforks just as easily if those labels were not already taken. But wave/particle is to existence as awareness/intention is to knowing. Both constitute an inseparable pair: dyads.

In describing the spectrum of consciousness, writers often use the term *dualism* to distinguish the psychological process involved in moving from one state to another. Often dualism is construed as an illusion, or in another sense, as a separation. A dualism distinguishes the nirvikalpa samadhi state from the existential state—the first dualism being the subject/object or the I/thou dualism. In the existential state all objects are perceived as separate; the connectedness of everything is not apparent. Whether the connectedness has been forgotten (sublimated), or never known is an important point. The Platonic Idealist would hold that all knowledge and perfect form already exist, and the human task is to become aware of this. In a mechanistic Newtonian universe this might be so, but in an evolving universe with entropic and negentropic processes proceeding simultaneously, all knowledge is not possible *a priori*; indeed, all knowledge does not exist.

The future always has yet to unfurl itself. It has not been created, and is unpredictable and unknowable. The past is knowable, but not necessarily through direct sentience. Were the past directly knowable, the ancients mystics could have known it as we know it today. But they couldn't. It has taken intuitive insight and rational thinking skills, together with scientific instruments, to decipher past realities. The past exists only as information that is stored in various ways—some yet to be discovered—and requires interpretation to give it meaning. The universe indeed exists as energy and is known by its patterns of energy, but interpreting the meaning of information is a function of the state of consciousness of the interpreter and the existing information base. The more evolved the mentality and the more information available, the richer the meaning attached to that information can be.

We can deduce that neither the existential state, the samadhi states, nor those in between, provide omniscience. An evolving macroscale universe is still creating itself through nonlinear dissipative processes sustained by underlying quantum processes. The samadhi states provide unequivocal feelings of ecstasy, peace, and a sense of the eternal that are vividly conjured even when not in the samadhi. The ancients interpreted these feelings as proof that the Creator assured eternal existence. But we can't be sure of the sources of these feelings. Are they but a shadowy memory from the security of the womb, or are they more profound? There's reason to believe they are an informational echo of the ground-state of being, the state of awareness of the primordial unstructured universe where nothing exists but potential.

The perception that all is Self in the ground-state is also widely interpreted as the perception of All That Is. The subjective experiences of ecstasy are likely primordial templates repeated throughout all levels of space-time organization. Why else would we have them? The sensation of eternity is taken as assurance by all sentient beings who experience samadhi that consciousness is an eternal phenomenon. The sensation of ecstasy may indeed be a memory from the womb, but even so, I would suggest that both ecstasy and the eternal sense are likely produced by the awareness of every cell of the body being coherently resonant with the timeless ground-state of being. After all, "feeling," or internal sensations, are just information to which the organism attaches meaning, in this case a pleasurable one. But this sensation is all-consuming, and affects the entire body; it is accomplished just by achieving a particular quiet state of mind.

This coherence, which has a possible mechanism in quantum physics called a *Bose-Einstein condensate*, could allow the entire structure of the organism to function as a coherent whole and to experience intense ecstasy and a strong feeling of security all at once.[2] The experience of the samadhi states for the mystic serves the same purpose that the discovery of the background radiation from the big bang serves for cosmologists: proof of the existence of the universe, and its origin and nature.

It confirms theory. In both cases, however, the experience just provides information to which the mind assigns meaning, and in both cases later data can invalidate the assigned meaning.

If we postulate that the experience of the nirvikalpa samadhi state is the experience of resonance with the ground-state of all matter, this would tie the Great Chain of Being back to its roots in the quantum potential of matter. The modern term for this energy field is called *zero-point energy* or *vacuum energy*. Vacuum energy is the presumed energy field that is in continuous dynamic exchange with matter, which sustains the form and existence of matter at the quantum level. To be aware of the zero-point is tantamount to being aware only of awareness itself.

Vacuum energy has been a subject of intrigue since the work of Nikola Tesla, Thomas Edison's rival, early in this century. Tesla, it is widely believed, transmitted energy through the atmosphere—quite possibly energy dwelling in the zero-point field, though no one has been able to replicate his efforts. But in recent years scientists have come to grips with the zero point as a reality. Zero-point energy is also interpreted as an infinite sea of unstructured energy potential from which the universe arose, which pervades all space in the universe (and probably outside as well). In this sense the universe did not arise ex-nihilo, but rather arose, was created, or "intended" from this underlying quantum potential. Because the field of zero-point energy is currently known only as a mathematical abstraction of wave forms that carry energy proportional to the frequency of the wave, matter itself may also be thought of in these abstract terms. What I would propose is occurring in nature is that the qualities of mind, which we humans experience as awareness and intention, and can trace down through the chain of organic and inorganic complexity, has its roots in the most basic field, the field of vacuum energy, and in the nonlocality attribute of the energy at this level. By this way of thinking, nature itself is in some sense aware and intentional.

23

In the early 1970s I became aware of the new proposal by British biologist James Lovelock that Earth and its ecosystem should be considered as an single organism rather than as separate and independent systems. His proposal was called the *Gaia hypothesis*. The concept struck a chord in me, as it gave voice to a new metaphor that I too could use.

In space the universe itself seems more organismic than divided parts, more intelligent than inanimate. But in the years immediately following the moonshot I was having difficulty expressing this notion. Extending Gaia to include the larger reality seemed a reasonable way to convey this idea, though the details of how this could be done were yet elusive. Intelligent organisms, after all, must evolve from something more basic.

In attempting to resolve the issue of how structure and order in the universe arose, the late physicist David Bohm proposed that beneath the level of quantum uncertainty lay an "implicate" order. Within this order is embedded the structure that emerges in the universe. His was an idea taken from Platonic Idealism that beauty and perfect form already exist, and need only to be manifested in the physical world. Bohm was looking for the physical mechanism by which form and order emerge in the macroworld. But this approach only delays without squarely addressing the tough question of what originally gives rise to order, indeed what initially gives rise to the laws of physics, and how nature became "lawful."

If the universe is fundamentally organismic, it isn't necessary for order to be present *a priori*. Rather, only the means need exist whereby structuring could be learned through trial and error. The universe doesn't need to know how to create order; it only has to have the capacity to learn to create order. Nonlocality provides a clue as to how order is maintained. And learning suggests the attribute we label as *consciousness*. Very simple combinations of quantized energy clearly shape themselves into molecules, and when combined with other molecules produce extraordinary structures—from ice crystals to the DNA double helix. If randomness were the basic impulse in the macro-world, processes would diverge, not converge toward greater order.

Chaos and complexity theory show us that complex atomic ordering is repeated at larger and smaller scales alike. But they are still deterministic theories. That is, if one starts at precisely the same place in the process, one will always get the same results. And they have also been shown to create deterministic self-organizing (autopoietic) systems. But we cannot imagine a way in which intentionality and creativity can arise from determinism.

If the universe is indeed a learning universe in which trial and error prevail, successful actions will accumulate and continue, whereas unsuccessful actions will fail and dissipate. A necessary consequence of this process is a means to evaluate outcomes, and with it the *perception* of dualisms must arise. Once the existence of physical objects comes into being, and self-organization continues toward ever more complex structure, some processes will successfully continue, while others will wither. Those that have successfully continued are those we label and map with the laws of physics, chemistry, biology, and cultural values. They have survived. We live among them every day. By the same token, unsuccessful processes may have left little or no evidence of their former existence. It's intriguing to consider what this universe would be like under slightly different initial circumstances.

If the universal processes were only slightly varied, and the physical constants slightly larger or smaller than the ones we know, the universe could not exist in its current form: atoms, stars, and galaxies could not have developed as they did. So it can be said that we live in a "successful" universe. Thus, *success* and *failure* are labels we may attach to the possible outcomes of any particular process, if we experience that process and have the ability to learn. Success and failure is not only a dualism but a dyad, two related, unavoidable possibilities of any intentional process. When a coin is flipped, it most likely will land on one of its two sides (with a vanishingly small probability of landing on an edge). The existential state of consciousness is said to be separated from the ego state by the awareness of death. The existential state of matter in the universe may either self-organize successfully, or not. Failure in this sense amounts to death. This separates one potential outcome for the existent structure from the alternative—nonexistence.

The point here is that opposite possible outcomes of any learning process form not only a dualism, but a dyad. The awareness that experiences the process, however, can only experience one or the other outcome—not both simultaneously. The two limiting outcomes of any process may be thought of as dyadic: success/failure, life/death, good/evil, pleasure/pain. Of course, it takes a successful, highly evolved, cognitive, and sentient system such as the one we possess to use the labels, to look back and make such an analysis. But less complex learning systems will still experience the processes and be subject to their outcomes. To the extent possible, it will choose outcomes that reflect success, yet the overall system will learn from the experiences of failure, death, evil, and pain. What the aware system has experienced will shape the meaning that is given to the experience.

The distinction between pain and pleasure might be the first that an aware organism makes. This is likely how it distinguishes between success and failure. A feedback learning process would cause self-organization to be immensely more efficient than random mutation, and more consistent with the rapid progression and discontinuities of the evolutionary processes we see in nature. Patterns of energy provide information. Information is an immediate consequence of nature's organization process, and is the engine that drives the processes of awareness, intentionality, learning, and mentality.

It's inconceivable to me that nature's processes might have failed to utilize information early in the self-organization of the universe in order to evolve the process of knowing. Whatever form the most primitive processes of awareness and intention took, they were likely destined to became more complex through learning. Is the awareness of a carbon atom more complex and "mental" than that of an electron? When hydrocarbons started organizing in Earth's primordial soup, awareness and intentionality certainly began to emerge on the scale that our instruments of today could detect if we looked carefully. Darwinian dogma simply loses credibility. Is the waggle dance of bees and the courting rituals of all species evidence of a subjective inner life? Aren't such behaviors intentional? Although we must be cautious in anthropomorphizing nature, for it seems certain not to be self-reflective at more simple levels of complexity. I wonder, however, if bees are as cautious about entomologizing us.

In a universe of natural process where awareness and intentionality exist, dyadic relationships are a fundamental necessity. Matter and consciousness seem to be an inseparable dyad, not a dualism with two irreconcilable realms.

If we postulate that dyads are a fundamental configuration in our mapping of nature, we can ask why it hasn't been discovered previously. One answer is our advantage of 20/20 hindsight through several millennia of self-reflective learning, and the lack of detailed information possessed by the ancients concerning the structure of the universe. A corollary answer is found in how our awareness manages information, the focus of our awareness, or "attention."

We can focus our attention broadly, as when looking at the heavens, or focus on the detail, as when looking through a microscope. We can focus our attention on the internal events of our awareness, or the external events perceived through the senses. But we cannot focus our attention on more than one of these areas at a time, though our awareness is still monitoring the information around its periphery. It would seem that the scope or "screen" of our attention can only contain a relatively fixed amount of information. That information can be in great detail, or all-encompassing. But the amount available to us at any moment is fairly constant. This is a function of *mentality*, the evolutionary complexity of the brain. The simpler the entities' awareness, the more limited the scope of attention. At low levels of organization, awareness and attention are likely the same.

Attention is curiously monadic, not dyadic. It is single-focused, not multi-focused; to attempt to simultaneously shift attention otherwise tends to cause disorientation. When we shift our attention from one object to another we become distracted. Indeed, one powerful form of meditation to create coherence and calmness involves focusing between things, between thoughts, and holding the attention on nothing. This method simply allows thought and images to pass through our awareness and fade away until the activity of the brain settles down to quiet nothingness. With practice, the method can lead to the samadhi state. Another method involves focusing continuously on one object without

From Outer Space to Inner Space

interruption. This causes the mind to discover detail it had not previously noticed. With practice, this method causes the sensation of merging with the object. Both are powerful tools that help discipline mental processes, reduce fragmentation and distractions, and enhance awareness.

Albert Einstein used a variation of the latter technique to improve focus and concentration while studying written material. He developed what he termed a "distraction index." By noting each time his attention was distracted from the material being read, he counted the distractions in a certain period of time. With each distraction he would affirm, "I won't be distracted again." With practice, he noted, the number of distractions was quickly reduced toward zero.

One consequence of our limited monocular focus of attention is that dualisms appear. But the dyadic relationship is missed. While focusing on one object, its connection to something else is not comprehended. When focusing on one face of a coin, the other face is not visible. When focusing on any object, the rest of the universe seems less real, or more removed, at that moment.

24

While the Institute of Noetic Sciences was still in its infancy and I was working with Brendan O'Regan, we frequently pondered the reported cases of extraordinary human functioning. What he and I discussed were the underlying mechanisms that made such "miracles" possible, hoping to find clues to our more normal capabilities.

On several occasions the subject was savants, those amazing individuals who are usually limited in normal thinking functions, but who can perform astounding feats in a particular realm, such as music or mathematics. Some are capable of instantaneously solving complex arithmetic problems—such as finding the cube root of an arbitrary 15-digit number. These functions don't take place in conscious awareness, but deep within the recesses of the brain's subconscious. When a question is posed, their answer pops to the surface of awareness like a thought, rather than as the result of a tedious mental calculation—often much more rapidly than a modern computer.

Utilizing such subconscious mentality doesn't require self-reflective awareness, but calls on a more primitive process that likely evolved much earlier: A question is posed to an individual, who responds by forming an "intention" to produce an answer. Shortly, a response is produced by the subconscious. Is the process different for more simple organisms? The bee doing its waggle dance hasn't devised instruments to measure direction and distance to the field of flowers, nor likely performs an intentional mental calculation. But it has stored certain clues, and it acts them out in the dance based on information stored in what we would label a unconscious or subconscious memory. Whatever the mechanism, there is awareness and an intent to store and to communicate information. The traditional explanation of a mechanistic "instinct" at work here just isn't sufficient.

In my studies I've witnessed experiments in which beans were grown in a controlled environment with the only variable being the kind and loving attention given by the experimenter. The "green thumb" effect clearly indicates that positive nonlocal intention promotes accelerated growth, and conversely, threatening thoughts and actions hinder growth. Undifferentiated awareness, even in plants with the most simple sensing systems and without the subject/object dualism, seems able to distinguish dyadic sensory input such as harm/benefit and pain/pleasure. But it is likely the evolution of multiple sensors for an organism and a means for managing larger amounts of information that led to the subject/object dualism and cognition of things other than Self.

The spectrum of consciousness model defines additional states below the existential by introducing additional dualisms or fragmentations of unified awareness. The state below the ego is labeled the *persona*, which is of therapeutic concern in humans. Oftentimes the characteristics of these states are accompanied by depression, fears, feelings of malaise, and various personality disorders. I am not addressing those conditions that are strictly chemical in origin, but rather those that result from emotional and psychological trauma (which may produce chemical pathology), or just deeply programmed beliefs that are out of step with the individual's changing environment. The latter condition probably includes most of humankind. The general therapeutic protocol includes recovery of sublimated experience, reinterpreting the significance or the cause of the experience, and integrating it into a new mosaic, resulting in healing the dualism. New meaning is given to past experience. Buoyed spirits, a heightened sense of well-being, and joy result.

But why should recovering lost or mislabeled information from the subconscious improve the sense of well-being? Why should happiness result from merely transcending a dualism in the mind?

The process is that of reorganizing information and intending something abstract we call *healing*. This is a completely subjective process. Happiness, joy, bliss, and their unpleasant counterparts are "feeling" states. They are states of existence that permeate the entire organism and are labeled by the mind as emotions. Psychobabble aside, well-being can only be improved by an internal subjective process if matter and the subjective aspect of mind are inextricably and dyadically linked. The only difference between influencing the body with a thought and influencing the outside world with a thought is the perceived separation, a local versus nonlocal process—but from the perspective of a cell in the toe, the brain is nonlocal.

Healing dualisms from the shadowy realm below the ego state is the business of the mental health professions. Healing dualisms from the ego state to the samadhi has traditionally been the business of religion—assisting the individual in transcending the physical world toward spiritual awareness. But both are increasingly in the business of "self help." The gradations of happiness, contentment, and ecstasy all improve as the dualisms are transcended toward samadhi. In a natural universe where things are knowable, this means integrating information to reveal a more holistic tapestry of knowing. Integrating information and deriving richer meaning in this universe leads to ecstasy. But it should lead to ecstasy only if that is the learned, successful outcome of a natural process, one that is purposefully in step with nature's processes. What other possible reason could there be for nature providing feelings of joy and malaise? Why should we have feelings at all? Were we but a deterministic collection of atoms, feelings and emotions would serve no purpose.

To attempt to integrate all the foregoing into a concept of reality that seems to describe the universe we experience in the early 21st century is a heady task. It has taken me 35 years to relinquish my preconceptions and put it all together in the form of a dyadic model; I hope to present it here in two pages.

For at least three decades, researchers have observed that individuals electromagnetically isolated in Faraday cages can achieve some sort of communication. Rhine, Schmidt, Puthoff and Targ, Zellerbatun, and others have all achieved positive results in this way. Others, notably Backster in the United States and laboratories in the Soviet Union, have demonstrated that lower-level organisms also seem to share nonlocal connection. In all traditional cultures of the world, folk medicine, nonlocal communication, universal interconnectedness, and mind-over-matter effects were the basis for cultural cosmologies and theories of existence. This was without exception until this century.

Today, however, the scientific work undertaken to understand these effects and the cultural lore itself is ignored, and disparaged by mainstream science on the basis that no scientific theory exists to support such phenomena. And any attempt to structure a cohesive theory is viewed as a threat to the cherished dogma of materialistic determinism. But what is tantamount to the missing link was found in 1982 with the Aspect experiment in physics. It demonstrated that basic matter is correlated nonlocally in ways that prevailing classical theory said couldn't happen.

At the limits of our everyday macroscale world, the measures by which we "know" how the universe works break down. At the sub-atomic level, structure disappears into dynamic exchange of energy with the zero-point field. Here nonlocality prevails. Space-time ceases to exist in this domain, as all exchanges of energy are reversible, continuous, and unpredictable. On a different scale size, but in a similar way, a cloud of water vapor hovering over a pond shares a dynamic exchange of water molecules with the pond. At the limit of the very, very large (the universe), we cannot imagine what is beyond—perhaps nothing but the zero-point field. Or perhaps the universe is infinite, and we just have not discovered the tools to help us know that. At the limit of the very, very fast (the speed of light), time becomes meaningless (in other words, stops) as matter becomes infinite, fills all space (wave characteristic), and becomes pure energy. At the limit of the very, very hot, all matter dissolves into high-temperature energy. At the limit of the very, very cold (supercooling), matter begins to behave coherently as though all molecules are nonlocally connected on the macroscale.

Our macroscale universe is a unique place, where matter and knowing exist inextricably together. Matter, space, time, and knowing do not exist anywhere else we can verify. The mathematical descriptions of nature all become discontinuous, imaginary, and meaningless, or result in singularities at the edges of our world—a result that distresses those physicists and mathematicians who still cling to the belief that all reality is ultimately mathematically elegant and simple. But the universe knows nothing of mathematics. It is simply a universe of process—a process we are attempting to know.

Energy, we know, is the foundation of all matter; information is the foundation of knowing. Both were present at the moment of creation, whether in a big bang, or in a continuous process of creation in galaxies. It is likely that just as energy produced the physical structure that we recognize as waves and particles in our macro-world, the seeds of consciousness were also present to produce awareness and intentionality. I suggest that these fundamental attributes of nature are dyadically coupled in our universe. They find their basis in the very ground of existence, the zero-point field, which exists outside space and time. The descriptions we give to both existence and knowing are defined only in the physical universe and have meaning only within the universe *and* within our minds. The zero-point field has no characteristics (that we can currently measure) other than energy, and patterns of energy (in other words, information), but energy with the seeds of learning: awareness and intentionality. It has no temporal or spatial signposts that have been measured; locality and nonlocality emerge from the same point. All points and all matter in the physical universe have their origin in and owe their existence to the zero-point field. It is ubiquitous, yet nowhere, simultaneously, providing the quantum potential for all physical structure and the basic potential for awareness and volition to exist. In a dyadic model of reality, existence and knowing are coupled, as are awareness and intention, and many other attributes of nature's processes. That is to say they are inextricably related.

At first glance the foregoing notion likely appears outlandish, just as it did to me a decade or so ago.[1] However, after several years of reflection it seems more reasonable than proposing a universe created by accident, or one created by an eternal, preexisting, omnipotent, anthropic deity (or deities). It also seems more reasonable than a universe in which the laws of physics were codified and stored *a priori*, with no explanation as to how nature "knows" how to obey the physical laws, or a Platonic universe where beauty and perfect form preexisted, or a many-worlds complex of universes that divide, or the wave function collapses, when we change our minds. It seems infinitely more reasonable to propose a universe where all physical structure and all mentality arose together, inextricably intertwined through the feedback process of learning.

Nothing preexisted but quantum potential and the potential to learn—the seeds of both existence and knowing. The most compelling reasons to suggest this model is that it results in a universe similar to the one we inhabit, and seems to correctly map learning processes in nature. Perhaps most importantly, it allows anomalous phenomena excluded by older models, and causes a number of paradoxes, such as Schrödinger's Cat Paradox, to disappear.

Portraits of Reality

Interpretation and Paradox

25

When I look up into the heavens today I oftentimes wonder just how
much of the universe we can ever explore, seemingly bound as we are
by the speed of light and existing energy sources. It would, after all,
take billions of years to reach the most distant galaxies and several thousands just to explore our own. In our time alone, however, we have
become a spacefaring civilization. So we dream the impossible dream,
hoping to make more sense of reality, and just possibly to discover that
the dream wasn't impossible after all.[1]

Our desire to explore is really our desire to reveal, and perhaps alter,
the structure of the universe. We secretly hope to find our place in the
vast scheme of things. This has been the larger purpose of every mission
into space; it is also the larger purpose of both science and religion.
Metaphorically speaking, we want to one day touch the face of God.

When we explore, whether the journey is into space or into the
nucleus of the atom, we construct mental models of what we have
found. Much of the scientific and mystical literature of our time speculates about the existence of a multidimensional universe, unseen spirit
realities, Many Worlds interpretations, and the like. What the scientist
is looking for is some kind of grand theory of everything, a theory that
encompasses the universe in one consistent symmetric mathematical
treatment that makes the universe knowable. A proposed multidimensional universe has been one result of the search. Speculations as to the
world disappearing when no one is looking is another.

If existence and knowing are equally fundamental faces of our
universe, as this dyadic model proposes, then existence and knowing
are interacting processes, not independent and absolute as previously
believed. In such a universe we will eventually discover whatever there is
to know about what we experience.

What is wholly amazing (yet the wonder of it rarely occurs to us) is our ability to imagine things that don't even exist. We can create mental images of unicorns and centaurs, we can harbor hopes and beliefs that may or may not be realized, we can dream dreams that are literally absurd yet metaphorically meaningful. This ability alone imparts a clue that we are more than just deterministic matter on the one hand or creative mind in an infinitely malleable universe on the other. We are, rather, something of both elements. The scientific theorist hopes that the Platonic ideal of beauty and perfect form will be revealed with mathematical simplicity and symmetry, where it lurks behind the messiness of the physical world.

The mystic, in his or her search, is simply looking for an explanation of where the spirit world goes when one opens the eyes, and finds the beauty, ecstasy, and eternity of the Creator in unseen dimensions that are accessible only through the mind. Considerable popular thought suggests that it is in these unseen realms where science and religion meet. The scientist, the futurist, the mystic, and the UFO chaser all look to other dimensions as a way to explain the unexplained.

There is no intrinsic reason, however, why the universe should be elegant, beautiful, simple, symmetric, ecstatic, or multiple, except that those qualities arise from the experiences, labels, and desires of the beholder. To a warthog, beauty rests in other warthogs; to a hammer, the whole world is the nail. The meaning of our experiences lies in the interpretation that we give them based on the memory of prior events and learning.

The multidimensional worlds of the scientist and the mystic seem to arise from different realities, but in truth the ideas of both arise from the same source—the creative thought process that seeks to *know*, and from the flawed assumption that mind and matter are completely different things. The creation of a thought, any thought, is merely the process of turning energy into information (in other words, patterns of energy), and then becoming aware of it. Yet it is a magical process, one that was taking place long before we gained self-reflective awareness. Unfortunately, most human thought is little more than static; turbulence in the tangled hierarchy. In the undisciplined brain, it amounts to a kind of idle chatter followed by emotional reactions. Turning the creative turbulence into coherent structure is the essence of knowing. A student of a meditative discipline can learn to quell the static and enhance all brain functions, particularly those believed to originate in the right hemisphere and the prelinguistic brain. Students trained in the thinking disciplines tend to enhance left-brain functions, which are dependent upon language. Those trained in both develop greatly enhanced benefits that add up to more than the mere sum of the two.

In either case, when the mind is stoked with new information, and questions are allowed to gestate for a time, additional creative thought surfaces in magical fashion to solve problems and express meaning by way of intuition. One does not consciously create a new thought, but rather ponders associated parcels of information that dwell in memory. Then, after a time, a new thought emerges into awareness from the process that organized the new information below the level of conscious awareness.

It's not surprising, by this way of thinking, that both the scientist and the mystic of old might have interpreted emerging new images and thoughts as literal, absolute, and as having arisen from the basic fabric of existence, due to the spontaneity with which thoughts arise. And in a sense they have. But by the same token, so have unicorns.

If I extrapolate on this dyadic model, it suggests that the tangled hierarchy of information developed through human history has not only increased our ability to consciously manage information and produce creative thoughts, but increased the number of possibilities available for giving meaning to ideas as well. That is to say, we have created additional ways to get completely off track. Only experimental validation can help decide whether our creative thoughts about nature's reality do in fact describe it accurately. And even if the description is valid in certain details, the larger meaning may still present difficulties and lead to dead ends, as has been the case in certain interpretations of quantum theory and mystical experience. In fact, knowledge can only progress through theories and interpretations that are falsifiable; that is, they must be framed so that evidence can demonstrate they are flawed. One must gain additional information and rise to a broader viewpoint to better see the vast and intricate terrain below. By doing so, one can discover and avoid the trap of dualism and dogma.

The issue of the multidimensional character of our universe is a fine case in point. Dimensionality involves both existence and knowing. This is at the heart of the scientific method: Measured quantities must have a dimension. Each basic measure, such as length, mass, or tune, are considered dimensions, and, as will be discussed later in the chapter, are often confused by the laity as spatial dimensions.

We experience ourselves as three-dimensional bodies in a universe of three spatial dimensions, primarily because we have a visual sensor, and three spatial dimensions actually do exist. Vision is the only sensor that directly senses a portion of the electromagnetic energies: light. Science has displayed convincing evidence that electromagnetic energy is nature's basic building block of physical reality; today all physical existence and knowing can be expressed as the organization and exchange of electromagnetic energy. Demonstrating that fact elegantly and simply is why physicists seek a grand theory of everything, as there are yet some very unseemly inconsistencies in the science.

But imagine if all humankind were blind and always had been. Had we not evolved a visual sense, yet possessed all the others, we might have discovered our world by believing it had five distinct dimensions, one each for touch, taste, smell, sound—and that mysterious internal feeling we associate with intuition. Without the visual sense and its direct clue to the existence of the electromagnetic spectrum, it would have been an even longer and more arduous process to discover that the basis of existence was energy manifest in electromagnetic form. We might still be trying to demonstrate that the smell, taste, and feel of an object could be unified and simplified with a grand theory of tasty objects. Furthermore, it would have been very difficult to reveal that our world really appears to consist of three spatial dimensions if the measures for smell, taste, feel and sound were first established as the most fundamental measures of things. It is significant that we remember from history that even with vision our early forebears believed that the basic elements were four: air, earth, fire, and water. Yet some still adamantly believe that intuitively derived information comes from an absolute source.

Our beliefs about the nature of existence are based on how we receive information, or stimuli, from the external world. We cannot be confident that our maps are validated until there are consistent theories of everything (meaning all basic processes). However, that doesn't mean all things must be in the same grand mathematical theory. Indeed it is doubtful they could be. Our species appears to have adequate sensors and information management capability for the task, but science has overlooked this mysterious inner sense, this nonlocal source of information, in its description of reality. Grand unification theories cannot be complete without understanding that information as well. It's possible that what the mystic experiences and has described for centuries as "chi" or "vital" energy may be something different that defies electromagnetic or quantum description, though I personally doubt that this is the case. Our experience of internal energy is likely a result of the way the molecules of the body interact with the zero-point (quantum) field. This is, however, a speculative idea that has yet to be verified.

However, both the scientist and the mystic can agree on what we mean by spatial dimension. This is the attribute by which we measure or know the extent of our world. The case in point is the space in which we exist, which is measured by length, breadth, and height—three spatial dimensions.

A point is said to have no dimensions. A straight line has one, a flat plane two, and a cube three. Mathematicians say any point on a line can be described by a single number. Any point in a plane can be described by two numbers, and any point in a cube can be described by three numbers. Curved spaces just require additional numbers to describe the curvature, and an additional dimension into which the space can bend. Both mathematician and mystic will agree that there are spaces beyond the three-dimensional space where things can exist. The physicist will agree in principle, but should insist upon experimental evidence that existence can in fact have access to and reside in the higher dimensional space. It's difficult to illustrate this point, as our brain cannot visualize more than three spatial dimensions—perhaps we cannot because they don't exist, or because we have yet to experience them. So we have to use a simpler thinking device that mathematicians have devised.

By simply laying out numbers in a square array like a chess board, it has been discovered that a space with an arbitrary number of "dimensions" may be mathematically described. This can be done by adding another row and column of numbers, properly positioned, to represent points in the next higher dimension.[2] Surprisingly, the entire array (matrix) of numbers may be treated as a single equation instead of several separate equations—a mathematically elegant and relatively simple way to describe a wide range of complex scientific problems. In its most basic form this is called *matrix algebra*, and in a more generalized form it's called *tensor analysis*. The trick is to arrange the numbers in the proper order and to describe the mathematical relationships between them.

A frequent point of confusion among nonscientists reading popular science accounts is due to the fact that all scientific measurements must have a "dimension." So, in addition to spatial dimensions, other quantities such as time, electrical charge, quantum properties, or what have you, are also considered "dimensions" of the mathematical description. A point of confusion arises, as there is an interpretation of the tensor mathematics that suggests that one can treat these numbers as though they represent actual spatial dimensions. This interpretation suggests that our universe is a multidimensional universe, and that the laws of physics result simply from the twisting and warping of multidimensional space. Though this geometric *interpretation* of the laws of physics is highly controversial, the elegance of the mathematics makes it very appealing. Use of the word *properties* or *attributes* instead of *dimensions* for nonspatial quantities might eliminate some confusion.

Einstein used such tensor mathematics to express general relativity, and it worked, although one of Einstein's "dimensions" was time, and of course not spatial.

Using this method, his equations correctly predicted, among other things, that light would bend around massive stellar objects, suggesting that space itself was curved, because photons, having no mass, should not be attracted to gravitating objects like stars.

Einstein's general theory of relativity is a geometric interpretation of the universe. In the relativity view, space and time are inextricably interrelated to show a four-dimensional space-time universe: three spatial dimensions and one time dimension. Certain predictions of the theory have been consistently validated for almost a century now, which has led others to create more complex geometries and topologies that might reveal even more mathematical elegance and simplicity for the laws of physics. Recent experiments in space, however, challenge the Einstein interpretation, and the completeness of the theory.

Theodor Kaluza added Maxwell's electromagnetic equations to Einstein's gravitational tensor in 1919, and demonstrated that Maxwell's and Einstein's theories could be joined and simplified by representing any point in the universe with a five-dimensional matrix. It took Einstein two full years of contemplation before he was convinced of its significance.

Only a few years earlier, all matter was discovered to be both wave and particle, and here were two major forces, gravitation and electromagnetic force, seeming to have a related interpretation—a bizarre and disturbing thought at the time for those still steeped in Newtonian reality. Later attempts have been and are still being made to unify all the laws of physics in this fashion. Success has yet to be achieved and validated.

To the mathematician working with higher-order geometries and not deeply concerned with physical interpretation, the rows and columns of the tensor represent actual physical dimensions. But the numbers are the same as those derived from gravitation, electromagnetic wave theory, and quantum mechanics. Whether spatial dimensions above three are actually required to exist in order to explain our world, or whether the metric tensor is just a sophisticated method of managing numbers, remains to be verified. Debate has been hot from both viewpoints in the scientific community, as physicists have attempted to bring all known physical laws into one multidimensional metric tensor. It's exciting and impressive, however, that the equations of physics do simplify by arranging themselves in such an array, replacing several equations with one that includes them all. Here is the elegance and simplicity that would exemplify Plato's perfect world. However, in our real world, all the terms are not symmetric and orderly, and therefore, still quite messy.

Dimensions of spatial existence are the same whether described in mathematical or mystical terms. The additional spatial dimensions of mathematics are just as abstract, etheric, and invisible as "spiritual" dimensions, and both are creations of mind as a means of knowing ourselves and our universe. Even if these other dimensions were actually to exist in physical reality, to be accessible they would require amounts of energy totally beyond the energy available on Earth now or for the foreseeable future, and/or be infinitesimal, on the order of nuclear radii. Therefore, even if we believe they exist, they aren't readily accessible to us by any known means, spiritual or physical.[3]

But there are other reasons to doubt the existence of dimensions or realms inhabitable by beings, but unavailable to us. One is that the universe seems to have evolved toward greater complexity, likely starting from the very simplest of origins. To begin with, a multidimensional universe begs the question of how it became so structured initially and then became so unstable as to collapse to the universe we observe, and an infinitesimal one that we can't access. But more importantly, general relativity predicts the bending of light around large celestial bodies, which it does do. To bend three-dimensional space requires a fourth dimension into which it can "bend." So the interpretation of this phenomenon has often been that the fourth dimension must, therefore, exist.

However, very recent work with zero-point energy theory also predicts the bending of light in the presence of celestial bodies. But not because our three-dimensional space is warped. The same elegant, metric equations apply in either case, but the physical interpretation of the meaning is different. Space doesn't have to be warped in order for light to bend around gravitating bodies. It just needs properties that influence electromagnetic propagation, which it has. Space is not empty, we now realize. The zero-point field doesn't lie in another dimension, but underlies the quantum structure of the macroscale world. It may be said to have no spatial dimensionality, or conversely, it may be said to fill all space. It is a reincarnation of the ether that was erroneously banished by the Michelson/Morley experiment to measure light in the late 19th century.

Another argument against multidimensional space is an evolutionary argument. Nature's evolution has produced species that must survive in an environment often hostile to life. All species evolve senses, shapes, colors, markings, and behaviors that permit survival in their environmental niche. Were we in a multidimensional universe and unable to sense that fact, our species would be at survival disadvantage. So the fact that we are unable to sense or even creatively visualize higher dimensions, except with logic and mathematical interpretation, speaks against their existence.

I conclude that unseen spatial dimensions are not likely available either to the scientist or the mystic as a verifiable explanation for macroworld phenomenon, in spite of the popular and scientific appeal of this idea. Nature and experiment, however, must be the final arbiter.

A dyadic model doesn't address unseen levels or dimensions, but it does suggest that the universe arose from the zero-point quantum field, rather than collapsed from a multidimensional one. Thus, it is more likely to be what it appears to be—a three-dimensional universe of physical and informational processes that require a number of interrelated descriptions, some of which can be gracefully simplified by using the metric tensor. The mathematical descriptions for relativity, quantum theory, and biology have indeed been validated in large measure; they accurately map the natural processes in certain domains. But at the edges of macroscale reality we run off the map of our current theories—and encounter the zero-point field. The map of the underlying vacuum energy (or zero-point field) is only now being constructed, though it is likely a rather simple one, made up of unstructured—but in some way correlated—quantum potential; a cauldron of unmanifest energy. Its properties offer a simpler and more physical interpretation of the world than crumpled and warped multidimensional space.

The zero-point field approach is particularly more appealing because variables such as time, mass, spatial dimensions, velocity, and momentum apparently have no meaning there, but may arise as a result of the interaction of matter with those energies. Only energy exchanges have meaning here—the most basic of concepts.

Time, in particular, is a holdover from Newtonian thinking when it is considered an absolute measure. Relativity demonstrates that it isn't absolute. Dissipative structure shows that it only moves forward on the macroscale of things; double-slit experiments demonstrate that time has no meaning at all in the subatomic realm. And special relativity demonstrates that time means nothing to a photon.

A paper published by Harold Puthoff and colleagues in 1994 proposes that even mass, gravitation, and inertia are not actually fundamental attributes of matter either, but may be viewed as the property of the interaction of the energy of matter with the energy of the zero-point field. This interaction may be thought of as similar to the resistance of a body moving through a fluid, but in this case it is energy moving through a field of energy. The theory, if validated, gives additional impetus to the zero-point as the source of nonlocality, and reduces the number of basic attributes of matter previously believed to exist. Recently, serious attention by a number of authors is devoted to examining the properties and potential of zero-point energy.

26

For at least 5,000 years human beings have suspected that there is something more to our being than just the physical body. Spirit is not so easily dismissed, as the strict materialist would have it. Yet if the spirit world does exist apart from the body and can produce creativity, thought, and other such functions, why do we need a brain? Indeed, why do we need a body at all? It seems rather doubtful that nature required 15 billion years, or some such time, to evolve a method of self-reflective knowing that can perform all the complex information management functions we find in a human brain/body, and also supports an independent, parallel system that's nonphysical.

Science has mapped most of the electromagnetic energy spectrum and structure. The energy density at the high frequency end of the electromagnetic spectrum is sufficient to permit a variety of undiscovered but usable energies. Perhaps these energies might possibly account for some mystical insights and visions. At least some think so—"higher vibes," you know. But in our space-time world, such high energies are destructive to macro-world structures. Frequencies above X-rays, and even prolonged exposure to sunlight or infrared radiation, damage our bodies. Atomic structures with atomic numbers greater than 92 are unstable and not found in nature. It is, therefore, quite difficult to imagine how nature might have organized energy or matter structures in the universe that we cannot detect, because we can detect the entire electromagnetic spectrum and the matter it composes and with which it interacts. Moreover, there is no known basis for any other fundamental form of energy. "Beings of Light," or angels, or even discarnate humans, as frequently reported in mystical writings through the ages, would emit the destructive radiation of a nuclear reactor, were they composed of such high-frequency energy. Wherever angels tread, deadly ruin would surely follow.

Again, the principle of parsimony suggests that realms, dimensions, energies, and entities not detectable from within the macroscale universe must be considered as metaphors for common experiences, which need a new interpretation based in modern knowledge. A dyadic model suggests that there are other interpretations when we consider the role of nonlocal information and the quantum level of nature's functioning: properties of nature that are particularly ephemeral and mysterious. The traditionally cited evidence and interpretations for the existence of such entities is more likely a product of limited knowledge, tangled thought, imagination, and a minimum of independent validation. The experiences are common enough and must be considered real in the sense that something significant happened. The interpretation, or meaning, may yet be obscure. A natural (as opposed to a supernatural) universe should be knowable and subject to validation. And, since the attributes we call *knowing* and *self-reflection* are demonstrably evolved attributes, eventually we should know the answers.

But how does the dyadic model explain the ubiquitous experience of persons of all cultures who claim to sense nonlocal information, who believe they have lived past lives, who experience unseen dimensions and entities, and seem capable of influencing matter nonlocally, as did Uri Geller, Norbu Chen, Sai Baba and others? Two seemingly unrelated concepts when taken together appear to offer answers: the concepts of holographic information and the zero-point field.

We've all seen the amazing display of laser technology that can produce three-dimensional images, though they are too ephemeral to experience with the touch. In other words, one can see them, but also walk through them. The three-dimensional hologram, made with coherent beams focused on a photographic plate, are common in any novelty store. More exotic versions are often projected in museums, stage shows, and so forth. Holographic technology is just an exotic extension of the infamous double-slit experiment, though it uses coherent laser beams to create an interference pattern, which in this case is a holographic image. The two-dimensional interference pattern contains all of the information necessary to reconstruct the image of the original object. The significant point is that a large amount of research suggests that the brain manages information in a similar fashion.

A hologram analogy is valid, not only for the visual sense, but for all the senses as well, it seems. The brain (and every cell of the body) is a quantum device. And every quantum entity has both a local (particle) and a nonlocal (wave) aspect. Prodigious amounts of information can be carried in this holographic manner, including, theoretically, the entire space-time history of the learning organism. Recent research, to be discussed in following pages, suggests that we, and every physical object, have a resonant holographic image associated with our physical existence. It is called a quantum hologram. One can think of this as a halo, or "light body" made up of tiny quantum emissions from every molecule and cell in the body. In other words, the totality of our physical and subjective experiences can be thought of as a multimedia hologram resonant with ourselves and the zero-point field.

We can observe the resonance of a violin string when a note is struck nearby corresponding to its natural frequency, or when a vocalist shatters a glass with a strong falsetto. But it's a bit more difficult to imagine every structure in the universe having a hologram and resonating with the underlying infinite unstructured energy of the zero-point field. This quantum-mechanical-type resonance is an exchange of energy with the zero-point field such that the "phase change" (interference pattern) of quantum emission carries complete information about the history of the system.[1] In other words, one may think of the quantum resonance as carrying the information to create a hologram of the entire experience of an individual, including inner experience. This phenomenon provides a possible explanation for a host of visions, apparitions, and encounters with other "beings." It also can explain Carl Jung's concept of the collective unconscious and the reason why archetypal symbols recur in dreams, regression therapy, and Shamanistic rituals, plus many other nonlocal mental events.[2]

Many scientists believe the brain is too hot to support quantum processes, but that view is rapidly giving way to evidence that not only is the brain/body a quantum device, operating below the level of classical processes, but also the brain is physical proof of a biological quantum computer.

The quantum hologram was discovered and experimentally validated by Professor Walter Schempp in Germany while he was working on improvements to magnetic resonance imaging (MRI) technology. He discovered that the well-known phenomenon of emission/reabsorption of energy by all physical objects at the quantum level carries information about the history of that physical object. The mathematical formalism used by Schempp is the same quantum formalism used in holography, thus the name *quantum hologram* (QH). Similar to interference patterns in laser holography, the information is carried in the phase relationships of the emitted quanta. An additional and important property of the QH is that it is *nonlocal*, meaning that it is not located in space-time, but, similar to our particles Princeton and Bangkok in the EPR experiment of Chapter 15, carries the information everywhere.

The quantum hologram is the first information phenomenon that directly links all macroscale matter with the quantum world. For most of the 20th century, quantum nonlocality was considered a vexing artifact of particle interactions, but with no application to the macroscale world. Nonlocal information was not considered usable. Further, it brought science uncomfortably close to the forbidden realm of mysticism. The last 10 years of the century, however, witnessed a major new interest in the area of quantum computing and nonlocal quantum effects, particularly in Europe, where significant advances have been made. It is now clear that in the proper circumstances, quantum information is both available and usable. The real issues that must be now resolved in order to better map these fascinating discoveries pertain to (1) the structure of the zero-point field that fills all space and supports all matter, and (2) the precise mechanism of nonlocal resonance that permits remote matter to share information in this way.

Given what is now known, certain research in progress, and with well-informed speculation as to what will be verified in the near future, I suggest that the following common experiences can be understood by means of the quantum hologram and related quantum phenomena: (1) nonlocal intuitive feelings about people, events, and objects; (2) all forms of psychic information (a form of intuition); (3) the therapy of past-life regression, and with it, the popular belief in reincarnation; and finally (4) the basic nature of our conscious experiences.

Were it physically possible to place a sentient observer within the realm of quantum fluctuations of the zero-point field, looking at the macroworld from beyond space-time, the observer should perceive the exchanges of energy in the underlying structure in electrodynamic balance and in resonance with Self, and experience the connectedness of all things. Those who explore the samadhi states report precisely this same phenomenon as scintillating points of light when experiencing the All-That-Is.

My own experiences in the meditative state confirm these accounts. But is this possible? Are our sensory mechanisms sensitive enough to detect these tiny exchanges of energy? A mysterious experiment we performed aboard *Apollo 14* on the way home from the moon demonstrated to me that they could be. Recent experiments also indicate that the retina of the human eye can in some cases detect single photons of light, clearly demonstrating quantum awareness. Nature indeed possesses the means.

While rotating in the barbecue mode as we sped toward Earth, Houston had us turn off the cabin lights and pull eye patches over our heads so that we could see no residual light. When Stu, Al, and I did so, and dark-adapted, we all observed something quite extraordinary. After waiting a few seconds, what looked like a meteor trail would flare by. These individual traces of solar particles in the form of gamma rays crossed through our eyes, stimulating an optical response, as they made their way through the vacuum of space.[3] The gamma particles had penetrated the command module walls, the eye patches, and our eyelids— even perhaps our skull—to register on the optic nerve. This demonstrated to me that we do have the means to register small packets of energy that interact with our bodies, down to the level of individual quanta.

I would suggest that noticing quantum exchanges while in deep meditation is precisely what the ascetic disciplines have been doing for centuries. The zero-point field resonates with matter at each point in the universe, but is itself outside space-time; it can only be described as infinite and eternal. When one shifts one's point of view from the samadhi to the existential state, where existence and location in space-time is the more prominent reality, then the zero-point field appears to exist at each point in the macroscale universe. One can observe from the zero-point or from the macroscale world. but not both simultaneously. In other words, by merely internally shifting one's point of view, one moves from the nonlocal god-viewpoint to the local human-viewpoint, yet need go nowhere. One's experience depends entirely upon the point of view that awareness chooses to experience within the spectrum of consciousness.

It is almost certain that science will find further refinements to the descriptions of zero-point properties and how better to use quantum concepts to map the experiences of conscious awareness. Nonlocal resonance with any part of the space-time universe should be attainable by shifting one's point of focus to the zero-point—for it resonates with all matter. Quantum holography as a carrier of nonlocal information is then available to the individual. Some individuals seem to make this shift of awareness more naturally and easily than others. They are able to consciously perceive nonlocal information, as their brain quells the noise and focuses upon the signal. But dogmatic beliefs to the contrary are certain to diminish the ability to perceive, as one can consciously choose to pay attention to external information and rational thoughts, rather than the subtle whispers of internal states.

The phenomenon of remote viewing, which anyone can accomplish with a bit of training, allows the explanation that the mind of the individual is in resonance with the nonlocal object of attention. Resonance is established in a number of ways: by asking one's subconscious a question, or by visualizing an associated object or some icon representing the target object. Clearly the tricky part is to bypass conscious thought and let the intuitive feeling emerge. In my own experiences, I have witnessed group efforts at locating archaeological finds, sunken ships, and a missing person. By allowing trained personnel to work independently, then combining their intuitive impressions, valid data emerges. I've also witnessed the efficacy of group prayer directed toward particular individuals. Traditionally this would be deemed miraculous, but I prefer to think of it as the result of resonance and human intentionality operating quite naturally.

I've found it intriguing to work with a number of skilled people who have routinely experienced both remote viewing and the so-called out-of-body experience (OBE). But I've noted a difference between the two phenomenon. In the out-of-body experience, there's a sensation of *being* there; the focus of attention is narrower. It's as though they were somehow peering through a narrow lens. This would indicate that more of the brain/body is involved in the resonance, and one's entire attention more narrowly directed. In the often-reported case of the individual on the operating table who senses that he or she is suddenly at the top of the room watching doctors work, the brain is transposing information in such a way that his or her point of view seems to be more distant and dissociated from the body. One's awareness is automatically detached from the fearful or painful ordeal and provides the sensation of bilocation. This is most likely a trick of the psyche, a transposition of an unpleasant point of view to a more distant and comfortable one. But the subjective impression is that the Self has left the body. It's quite likely that this is an ancient mechanism of nature to manage trauma in an evolving world dominated by a brutish prey–predator food chain. This interpretation of the OBE is also quite consistent with the holographic way the brain manages information.

There is a very practical way one can validate this interpretation and also use the knowledge in everyday experience. To deal with minor injuries and accidents, one can immediately recreate the scene, allowing the body and mind to directly confront the painful circumstance. Alternatively, one can allow the mind to totally detach from the circumstance, sometimes producing the sensation of going out of the body. I've used the confrontational process with my children and myself hundreds of times to deal with the normal scrapes and sprains of daily life by simply reconnecting the injury with the offending object and focusing awareness on the point of pain until it subsides. This usually takes a matter of seconds. Neurophysiologists have explained to me that the procedure causes the release of a set of endorphins that relieve the painful sensation and allow healing to begin. And I'm sure this is true. But basically it has to do with controlling one's point of awareness and taking responsibility for controlling the pain. The autonomic nervous system then does the rest.

The classic near-death experience seems to be but an extension of the OBE, but carries overtones of the samadhi along with the emergence of archetypal images that provide assurances of well-being and eternal survival. That near-death experiences occur in all cultures, and that the entities that assure and assist the individual are entities from the religious lore of that culture, is telling. I would interpret such images as archetypal information recovered either from the deep subconscious or from nonlocal memory, rather than the appearances of discarnate entities from other realms.

Work done by Stephen LaBerge and his associates in the field of lucid dreaming also suggests additional mechanisms. The lucid dream is a merging of conscious awareness with the dream state, but without the restrictions of external physical reality. While in the lucid dream state the individual can learn to influence the drama, its course, and therefore its outcome.

In Tibetan Buddhist tradition, the lucid dream is an important discipline in preparing for what is called *bardo*, the transition from this life. One can interpret the classic near-death experience as a lucid dream in which the individual's desire for assurances and a sense of the eternal are acted out. For those individuals who are inexperienced with the samadhi states, the near-death experience is a spontaneous introduction to those states of consciousness, but uses images and archetypes that are already meaningful to the individual. Again, these archetypal images are derived from the religious lore of their particular culture. I have experimented with each of these techniques (except the near-death experience, of course) countless times in order to confirm for myself what others report and to personally experience the sensations. With a bit of training anyone can glimpse these nonlocal perceptions.

The notion of reincarnation may also be viewed in the dyadic model as an experience of nonlocality. The experiences of any life, both internal and external, may be thought of as information; any event before "now" is just information stored in memory. The future doesn't yet exist. It is presently but a hope or dream.

All we can experience is the "now," and only the now. Whatever time the clock indicates, and whatever the state of our consciousness, now is the only real time that exists for us. It isn't yet clear how detailed the storage of experience is, the reasons why some experiences are stored and others of significance are not, nor exactly all the modes and means of storage, particularly through nonlocal means. Certainly there is evidence that suggests that memory resides not only in the brain, but is also associated with our every cell, local and nonlocal memory alike. Rupert Sheldrake has explored the memory issue, and uses the terms *morphic resonance* and *morphogenic field* to suggest how information is stored and recovered in nature.

This approach is clearly in the right direction. Sheldrake initially suggested that the morphogenetic field may be species dependent, but subsequent work broadens the application. I would suggest that a more evolved species such as ours may have access to information of the simpler organism as well. We can form the conscious intent to resonate with an animal or an object, and the brain/mind's holographic information processes responds to the intent. An example of this are people such as my father, who have exceptional rapport with animals. Certain folk are demonstrably able to improve a creature's well-being and performance by intuitively diagnosing and relieving trauma unrecognized by less sensitive individuals, as in the case of certain well-noted animal psychologists. Subsequent Sheldrake work reports studies of dogs who intuitively sense the approach of their master well before his or her physical appearance. I have also observed the phenomenon in horses while the master is still approaching from several miles distant in an automobile.

It's this access to nonlocal information that the dyadic model suggests is the key to the experience of reincarnation. Ian Stevenson's scholarly work in this field clearly demonstrates that many living individuals have verifiable memories of a past life, particularly in those cultures that have a religious tradition incorporating reincarnation. In whatever manner information representing life's experiences is preserved, such as the quantum hologram, it is available to subsequent individuals who are able to access the information through nonlocal means. Memories of past lives seem to include subjective impressions as well as external events, indicating that both sensory information and the emotional interpretation of the information is retained. From the point of view of the living individual, all events before "now" are just memory. Yet these memories, which seem to emanate from before birth, must somehow be rationalized by the brain. The traditional explanation is that the particular "soul" lived in a earlier body.

The reincarnation hypothesis is a rational explanation within the East Asian belief systems. Stevenson's work clearly indicates that many of the cases he has investigated were verified, but it only validates that the information was nonlocally perceived. It doesn't validate the presence of "soul" or a notion of discarnate existence, which are only interpretations of how the phenomenon is possible. A dyadic model better explains why a number of people can claim to be a reincarnation of the same ancient individual. It better explains why, historically, famous people—Cleopatra, Joan of Arc, and other renowned figures—seem to "reincarnate" in several contemporary individuals: The current individual is recovering the memories from a nonlocal source.

Psychological "regression" is often claimed as valid evidence for both the existence of the soul and the reincarnation hypothesis. Regression in actual time does not really take place. The individual merely recovers information from memory and recalls it "now." The individual is already self-aware and likely self-reflectively aware, if he or she is post-puberty. The levels of evolving awareness are irreversible processes, as one cannot unlearn self-awareness. When one regresses under hypnosis to childhood, or even prenatally, there is the clear feeling of being at that time and place. However, this is a deceptive feeling, as one still possesses all the knowledge and rational faculties of the adult. Information stored prenatally, before self-awareness develops. which is then recalled after that event, does not validate either prenatal self-awareness or the existence of a soul. The memory of time spent between incarnations, which some claim, has no means of validation either.

The objective of therapeutic regression to a "past life" is to recover (or even create) information that gives new meaning to an old trauma. The meanings attached to the tangled hierarchy of information must be recalled, re-sorted, and reassigned in order to provide a more coherent, consistent, and rational understanding of Self. Regression therapy consistently demonstrates that providing new and a more highly integrated subjective understanding of past traumas, even if not externally verifiable, promotes a greater sense of well-being.[4]

27

For individuals who have experienced samadhi, out-of-body events, or near-death experiences, their lives are never again the same. The mystical literature, and now the popular literature, is full of such accounts. The desire to live life to its fullest, to acquire more knowledge, to abandon the economic treadmill, are all typical reactions to these experiences in altered states of consciousness. The previous fear of death is typically quelled. If the individual generally remains thereafter in the existential state of awareness, the deep internal feeling of eternity is quite profound and unshakable. But for those who haven't had such experiences, there's usually a greater anxiety and doubt concerning what happens next after this life, even if they profess belief in traditional religion.

Most of us live our lives in the ego state, that level of consciousness where the eternal is lost. The ego can also be defined as that portrait of ourselves we present to others, and sometimes to ourselves, which we believe defines who we are. The external presentation is usually just a mask or a caricature of the larger subjective reality underneath. When we think of an afterlife, it's this limiting internal sense of self-identity that we hope survives. We want to be the same individual entity in the next life as we are in this one—but without the problems. This is who we believe we really are, this is the picture we believe speaks of us. However, those who have healed, or transcended the ego dualism are certain that whatever form it takes, survival of our essence is assured. The precise mechanism is an object of passing curiosity, but not of fundamental importance; there is a trust in the process, whatever the process is. In the dyadic model, life experience (information) is not lost, but carried forward in the quantum hologram.

Death is merely the way nature renews itself and allows the creativity of the evolutionary process to proceed. In a dyadic model, even subatomic particles may be said to "die" as they give up their packet of energy and individual existence to transform into other forms of matter, which they routinely do. This is necessary in order for both the entropic and negentropic processes to proceed. Not only energy, but also its quantum attributes are transferred in the interaction, and the history of the interaction is carried forward. Thus the cycle of birth and death are patterned at the most fundamental levels of existence, and repeated at all scale sizes. Yet all is conserved and nothing ever lost.

Learning and experience in particular are not lost when an organism or particle dies. In traditional theology this is considered the domain of the soul for humans, though the underlying process seems to be patterned after the continuous cycling of energy into new forms of information. In a dyadic model, the human soul is the coherent structure of information experienced throughout one's life, and is carried in the quantum hologram. It is much more than just the ego structure. However, in most classical belief systems the soul includes discarnate existences of the ego, and a belief in other dimensions within which it resides. The religious traditions also provide the discarnate with all the incarnate attributes of sensory mechanisms and thinking capability. But because the human seems little different from other animals, apart from our more evolved self-reflective capability, it is unlikely that our postmortem information is handled much differently than theirs.

The idea of *soul* has made an impact on those in the scientific community who theorize how the distant future will look. In what will surely become a classic book, *The Physics of Immortality*, cosmologist Frank Tipler speculates that human beings will likely expand their presence throughout the solar system, the galaxy, and eventually the entire universe. He argues, though from a materialist perspective, that information is not lost in death, and could be captured and replayed on the extraordinarily powerful computers of the future. Using the strong artificial intelligence argument, he believes that storing information on these giant computers could perfectly emulate the human of today, and that the internal experiences of the virtual world would not be different from those of an existing human being.

Though his argument is ambitious in scope, it's fundamentally flawed simply because our subjective experiences are our *only* authentic experiences, and because the nonlocal characteristics and resonances of a system of computer chips will most certainly be different from those of a biologically evolved organism. There may be evolved and intelligent organisms in the universe that use a different natural biology than our own hydrocarbon biology. It's conceivable, but unlikely, that silicon is one of the basic ingredients in an anthropic evolutionary anatomy. Also, in allowing his humans and computers of the future to gain control of the physical universe, Tipler slips silently across the epiphenomenalist line in the sand by allowing a mechanistic system, which is inherently deterministic, to gain control of the process from which it arose. But so have the interpreters of quantum mechanics. It is the subjective experience that causes living systems to be markedly different from machines, not the energy processes in the external world.

In my opinion our best bet in assuring the eventual survival of our species is similar to the scenario Tipler suggests. But rather than counting on bionic computers, I find it more promising were we to allow the natural process of evolution to continue, aided and augmented by an informed, kindly, and intentional science that fully understands the processes of consciousness. In a dyadic scenario, the bionic processes remain the assistant, not the destiny.

The notion of personal survival that comes from a dyadic informational model is that humans of the future, greatly expanded in their natural capabilities, will be able to access information from the past that represents individual lives. It will be impossible, even in principle, for such a future individual to distinguish whether he or she is an "old soul" or just a very aware, intelligent "new soul" with memories of prior life experiences. As volitional human beings, we can engage in practices now that will assist in having our experiences and understanding carried forward to future generations.

The Tibetan Buddhist notion that our lives today should be largely devoted to preparation for dying is precisely focused on this very point: to assure that the wisdom of today is "reincarnated" and carried forward to the next generation of aware and thinking matter. The final act in this process of dying is selecting the proper individual who is about to be born, through nonlocal resonance and lucid dreaming. In this way the experienced wisdom of the prior life is made available, keeping the chain of evolved knowledge unbroken.

A fundamental prediction of a dyadic informational model is that the evolving, intelligent awareness that resides in the individual human being will continue to evolve toward the god state. Earlier humans, not comprehending a continuously evolving universe, believed that the omniscient god state already existed, and invested omnipotence and primal cause in such a preexisting being (or beings). A dyadic model predicts that evolution is ongoing but coming under conscious control, and the responsibility for the success of human evolution rests in the conscious choices made by and for physical hands.

To suggest that evolution is coming under conscious control also implies that it has been under subconscious or unconscious control—that is, without the self-reflective intellectualizing. Witness the thousands of living species, in some sense aware and perceptive, finding their niche in nature without self-reflection. The mystics have always suggested that transcending "thought" is the key to other states of consciousness. Modern evidence agrees that expanding awareness, through meditation, hypnosis, and holotropic breathing, in order to perceive at more subtle levels, stimulates conscious awareness of suppressed internal and nonlocal information. When it is said that we "create" our own reality, it is largely at the subconscious level that the action takes place. Therefore, expanding conscious awareness to include functions submerged in the subconscious must be part of our larger evolutionary process—else we would be unable to do so.

But surely nature hasn't placed all its eggs in one basket. Whether we humans accept responsibility for our lives and the destiny of our civilization, and barely survive, survive abundantly, or become a cosmic failure—which seems likely if we fail to develop greater awareness—is of concern mostly to ourselves. Surely other flowers have grown in the universe, should we turn out to be weeds. In a learning universe there must be failed processes; whether ours is a success or a failure is our own collective choice.

The individual who exhibits startling conscious control with mind-overmatter processes represents only the tip of an iceberg in an aware and intentional sea. The untapped potential that lies just under the surface is almost incomprehensible at our current stage of evolution. The continuously recurring appearance of titillating, spooky, and misunderstood events such as poltergeist effects, apparitions, miracles, stigmata, hauntings, and the like, are all of the same nature and quite easily explained within a dyadic model. They are all the result of human intentionality at either the conscious or subconscious level, working within quantum processes and the zero-point field. And they foreshadow greater conscious control over these capabilities.

Such events cannot be dismissed as medieval superstitious non-sense, though the medieval and classical interpretations of them are clearly flawed. It should be no less amazing that a singer can break a glass with her voice across a continent using high-fidelity transmission equipment and unseen waves, than that an angry adolescent, afraid to express himself to his parents, can cause physical objects to fly around a room by nonlocal means. Nor is the phenomenon particularly unlike an ardent child who can evoke holographic images of the Virgin Mary, or a grieving father who can perceive and be touched by the apparition of a deceased daughter who assures him that she is safe. Such is the power of intention in an aware, self-organizing universe that is nonlo-cally interconnected and has access to nonlocal memory stored in the zero-point field. These events are no more amazing or mystical than the phenomenon of a creative thought spontaneously arising in the mind. We all experience the latter and accept it as natural, but not all have experienced the former, and thus it seems bizarre. But we must keep in mind that life itself is a mystical experience of consciousness; it's just that we have grown used to it through the millennia. Within our evolving human potential are the very attributes the ancients attributed to the gods.

In a dyadic universe, the traditional Western idea of *final* judgment with the consequences of heaven or hell simply does not hold water. We live in a continuously evolving universe; the processes by which it operates are perpetual. Heaven and hell both may be said to reside within the living, conscious experiences of the individual. However, a frequently reported characteristic of near-death experiences is a seemingly instanta-neous review of life's choices in the larger context of their effect on the Self and others. In a sense, this is a (near) final evaluation by the Self of life's experiences. If this event in fact does take place at the point of actual death (which, of course, we can't verify), it would be consistent with a notion of final judgment and with the dyadic model of the evolution of information.

The notion of karma, interpreted as cause and effect, is also consistent with the dyadic model. If the existing individual is accessing "prior life" information, even subconsciously, it will create effects in behaviors, and even in physiology, in the same way that beliefs, choices, and experiences of this life create effects at the subconscious level. Therapeutic "past-life" regression has consistently shown that retrieving traumatic past experiences and reassigning new meaning and context to the "memory" relieves the trauma, though the event may not be verifiable in any physical way.

However, those interpretations of karma which require an external judge or judgment are not consistent with a dyadic model. In a universe such as ours, learning evolves and experiences accumulate, creating broader understanding, individually and collectively, of moral and ethical issues. Causality is vested both in the stability of ongoing natural processes and in intentionality. Consequence also lies in both.

In a connected, volitional universe, what we do to others we do to ourselves, as the transcendent Self of the samadhi is the shared experience for all. If we deny intentionality for the Self, we deny personal responsibility. By acknowledging intentionality as a causal factor, not only is personal responsibility a consequence, but also ignorance can only be a mitigating circumstance—never an excuse. Accepting intentionality as causal inevitably spawns personal responsibility. The de facto history of human moral development is precisely in keeping with a dyadic model: learning and improving through success and failure, slowly but repeatedly through the centuries to improve social well-being. Consequence is gradual and a result of the larger process of nature's learning. As Friedrich von Logau observed in the 17th century, though the mills of God (read: evolution) grind slowly, they grind exceedingly small.

Traditional thought structures, however (determinism in the case of science, and causal deity in most religious traditions), are a welter of confusion and contradiction concerning the origin and meaning of these issues. Consequence is either in the hereafter or in secular political decree. In existing belief systems, responsibility is imposed by authority, ecclesiastic or secular, not assumed by individuals as a consequence of learning and natural process. But it does not appear that the threat of retribution is a deterrent to immoral behavior. Only *enlightened self-interest* seems to consistently produce morality. The learning process isn't flawed; it is, however, hindered when we attribute causality to mechanistic determinism, or to remote, divine entities. We have to look across centuries to see the greater trajectory of the learning process. The ancient indigenous wisdom, to live in harmony with nature, was certainly more responsive to the idea of personal responsibility.

And what of the gentler experiences?

Love, goodness, kindness, and gentle demeanor have been virtues advocated by the esoteric traditions for at least three millennia. But the esoteric practice of those virtues has lagged far behind. The Christian faith emphasizes the loving nature of God and offers that as an example for humans to follow. But are love, goodness, and kindness really embedded in the structure of the universe? Why are they present in the first place, and how did they arise?

The gentle attributes are all correlated with the subjective experience of "feeling good." By making them virtues we codify these experiences in the cultural ethic. Clearly they're an internal signal as to the health of the body. If pleasant feelings are with us as a result of virtuous and kindly acts, or even virtuous and kindly thoughts, we have to ask why this is so.

A dyadic model suggests that all such feeling experiences, good and bad, are the learned behaviors of an evolutionary organism. All positive or good feelings, by whatever label, are a nuance, a shadow or shade of the ecstasy of the samadhi experience, which is itself a primordial sensation indicative of successful outcomes. All negative or painful feelings are precisely the dyadic obverse. If the model is valid, then good feelings are an indicator of successful thought and behavior, and should convey teleological advantage.

However, the latter-day evolution of self-reflective thought, tangled hierarchies of information, and a wider range of intentional actions, often pits rational choices and actions against "feeling" choices. The sexual experience is a basic example. The sensations of sexual coupling are nearly those of the samadhi experience, but unmanaged sexual activity leads to disastrous personal and social consequences. Witness the world's current population dilemma. Even the samadhi experience itself must be managed in light of personal and social needs. The lone hermit seeking only meditation and enlightenment must still chop wood and carry water, and likely makes minimal contribution to an evolving society, as his intentionality is mostly inwardly directed. Thus it would appear that our rational thinking must help in choosing between short-term and long-term pleasures. Here is another indication that evolution is coming under conscious control.

In esoteric and New Age thought it is often suggested that we follow our "hearts." This is a euphemism for the feeling nature of the body and the subconscious. In ancient lore the heart was the seat of the senses, not the brain. As I've already suggested, following the feeling nature alone is a mixed blessing. Because the feeling sensations respond not only to conscious thought, but to subconscious activity, internal health, and nonlocal resonance, it's vital that the individual becomes sufficiently aware of the differences in these inner signals to understand the source of the information if it is to be used constructively. That, of course, is what the meditative disciplines purport to provide—increased inner awareness and experience to lend appropriate meaning to one's subjective signals. From whatever discipline, the admonition "know thyself" is good advice.

Perhaps the most misused and overworked word in the modern English language is *love*. It's used in so many ways—from passages in scripture through pop songs to manuals on child-rearing, and as a euphemism for sexual activity—as to have lost any specific meaning. Perhaps the most unfortunate usage is as a mask for unrecognized dependent and codependent behavior in virtually every kind of relationship. Although the medieval Arthurian lore of chivalry and romantic love served a noble purpose in lifting womanhood from the barbarism of the Dark Ages, it also set the stage in the Western world for trivializing the conscious skills required to maintain harmonious love relationships. It has caused us to believe that love is something that comes unbidden and is in some way a universal absolute. The only time I am sure the intended meaning of the word is unequivocal is when it is accompanied by a spontaneous hug from a child who has yet to have his or her understanding of love corrupted. In its loftiest sense, love derives from a heady feeling of ecstasy toward the object of the affection, and results in responsible and empathetic attention. Curiously enough, love is only experienced in the absence of fear. The ecstasy of the samadhi is the prototypical love experience, in which nature sets the tone for how love should be experienced.

An ancient saying from the *Tao Te Ching* places these issues in proper perspective:

When the Tao is lost there is goodness,

When goodness is lost there is kindness,

When kindness is lost there is justice,

And when justice is lost there is ritual.

Ritual is the husk of faith and hope

and the beginning of chaos.

The Tao is interpreted as *the way*, or as understanding. The saying does not mention love, nor ecstasy, nor any other subjective sensation, except faith and hope—at the bottom of the list. Though the saying emphasizes behaviors, clearly those who can exhibit such desirable conduct must experience the inner peace and the upper range of positive feeling that are meant to be described with the words *love*, *bliss*, *ecstasy*, or what have you. The qualities proposed require the thought process as well as mystical understanding; that is to say, the tangled hierarchy of information, both mystical and rational, must be well sorted, assigned meaning, and understood. Yet the Tao is only experienced.

Interesting anomalies appear in nature from time to time that speak of her ingenuity. Such oddities underscore her ability to adapt and survive. Take, for instance, individuals with sensory impairment. They too construct internal maps that allow them to function in the external world quite well, though with difficulty. They typically enhance the sensitivity of one sense to compensate for the loss of another: the blind who read with their hands (without braille), or those amazing individuals who can "see" sound or "taste" visual images. Their brains process subjective information quite differently than most, yet allow them to function acceptably—perhaps in some cases even better.

One can deduce from this several ideas that are important to our understanding of consciousness processes. First, there is a significant difference between external and internal reality as the sensors of the human organism don't possess sufficient range to map all external reality, and require the extended sensing of modern instruments. Yet it seems we can sense all the various forms information takes in our macro-world.

Whatever external reality is, our understanding of it is only arrived at through a torrent of information and the process of assigning meaning to it. This requires communication and the exchange of information so that individuals may arrive at a consensus about external experience. But we cannot directly compare our experiences, only our internal maps of those experiences, as there is no objective map of reality, only consensus about experience. The evolutionary complexity of knowing and self-reflection evolved in parallel with the complex processes of physical evolution. If this is true, it seems all informational processes should be available to us, as they convey survival advantage. With creative thought we can come to know these processes with just the senses nature provided, and the sensory extensions we have created.

The language process, both written and oral, is a linear, left-brain function utilizing symbols. While quite specific, it is notoriously slow and inefficient in expressing our internal map of reality. Art and visual imagery are not only more efficient, but they also speak more directly to the prelinguistic portions of the brain that provide direct access to the subconscious and to the archetypal imagery stored from our ancient, ancestral past. They evoke feeling reactions first, not thought. "Knowing" is not just calculation, as is often believed in scientific circles, particularly in the artificial intelligence community. It involves a vast and tangled hierarchy of information derived from the senses, both local and nonlocal, and from thought, to which meaning is attached. Knowing is also accompanied by a feeling of certainty and experience that involves the entire organism, not just the thinking portions of the brain. Because of these subjective experiences, I have to believe the likelihood is remote that bionic machines can ever be made to emulate humans. Whatever the internal experience of a computer chip, it most certainly will be different from that of a natural organism.

Synthesis

Nearly 40 years ago I began to construct this dyadic model without presuming what it would evolve into. I held no assumptions other than that the universe manifests itself through natural processes that are knowable. If that assumption were flawed, I would have arrived at a dead end and been required to accept supernatural process. After almost four decades of sifting massive amounts of data, old science, amazing new science, traditional beliefs, and interpretations, no dead end has yet appeared.

I allowed that the experiments and mathematics of science were valid, and the inner experiences of the mystic were as well. However, both science and mysticism are rooted in language, thought structure, and political history, all of which were set in place long before we learned that our universe possessed evolutionary processes. Assumptions, interpretation, tradition, and dogma have accumulated through the centuries, along with paradox and contradiction, layered like cobwebs in the belfry of our collective thought. And it is these cobwebs that dictate much of our understanding today.

When I founded the Institute of Noetic Sciences in 1972 to pursue these studies, I accepted the struggles and triumphs, the encouragement of faithful friends, and the disparagement of detractors as a part of a larger process. Daily life itself, I came to recognize, was the only vehicle for learning, as each life expresses one of nature's emergent potentials that may prove significant for the whole. And it was in this way that I came to trust the process as synchronicities occurred time and again to destroy a meticulously constructed plan and replace it with a bold new direction not previously apparent. Intuitive feeling soon came to be a trusted friend, with rational thought and planning used to fill in the details.

Notions of cosmic accident and omnipotent design were set aside as questionable sources for these surprising synchronous events. Our capacity to survive and to accumulate knowledge in this fashion, and exercise volition, would not be possible unless the seeds for their existence were somehow rooted in the fundamental structure of the universe. Other universes may have come and gone, expanded and collapsed, tried and failed, because they had characteristics different from ours. But we'll likely never know.

Our own universe has survived long enough to produce sentient beings who seem to possess volition and can learn. And volition, however limited, is sufficient to eliminate determinism in its strictest sense. As late as the early 1900s there was no way to credibly falsify the materialist argument that volition is an illusion and that we are merely inhabitants of a universe grinding mechanistically and inexorably toward our predetermined end—and not under remote supernatural supervision. There was no way to convincingly argue that life was not a film strip in which the next frame would reveal the consequences of the prior frame's choice. Quantum mechanics opened the door for choice among the possibilities permitted by the laws of physics, before it stumbled against the entrenched Cartesian barrier of mind.

Existence requires matter and energy, and knowing requires information. All of these components of physicality were present from the first moments of galactic creation. Not until I realized that volition requires perception, and I'd made the assumption that it was omnipresent in some form, did the dyadic model begin to look truly interesting. Knowing and learning must have evolved right along with physical structure. As I worked through the various issues, paradox dissolved and descriptions of reality simplified. All I had to do was take proper care in understanding the interaction between what exists and how we know—a study in epistemology. By 1987 the model seemed solidly in place in my own mind, and I was startled by how it seemed to depict the universe we experience.

I didn't start out with the idea of dyads in mind, but along the way the presence of dualisms kept appearing: wave/particle, positive/negative charge, mind/matter, body/spirit, religion/science, male/female, yin/yang, and what have you. The dualisms all related either to existence, knowing, or both—all of which are complementary and/or opposites.

This seemed to me rather fundamental. Dualisms that cannot be separated are better described as dyads, the opposite faces of the same coin. Dualisms result when an observer gets stuck observing one face or the other and does not transcend to a broader viewpoint.

Then something else occurred to me, the paradox of all paradoxes: Determinism and volition result from dyadic considerations as well. Volition is an attribute of existence; determinism a product of knowing. Volition is required for creativity and change, and determinism is a result of discovering stability and continuity in nature, and then becoming fixated on stable, continuous processes. Creativity and stability must exist side by side in our universe, as they too are a dyadic pair. Unlimited creativity would be totally chaotic, but strict determinism would void innovation, the cornerstone of evolution.[1] Science is not destroyed by the presence of volition in the universe, just as mysticism is not destroyed by the presence of science. Both are just faces of knowing.

But there was something distressing about the results of a dyadic, evolving. learning model for the universe: In such a place there are no immutable guideposts for fallible humans to follow, no submerged absolute laws. Only trial and error and learning based upon experience. However, the pre-sentience universe has had several billion years of experience to its credit, and has learned successful behaviors in that time. Both science and religion codify them as laws. Civility, kindness, and love arose as choices to offset brutality, violence, and powerlessness in our more primitive social milieu. The inanimate universe is violent and explosive on all scales, and the early animate world only slightly less brutal, where only the fittest survived. Our earliest gods were angry, violent, and capricious, because that was the way the world appeared to our ancestors. Much later, we invested our gods with wisdom, love, and caring, because that was the essence of the world we wanted to live in; that was what we wanted our world to become. So arises an important point: we will ultimately become, both individually and as a civilization, the essence of our deepest intent.

A few years ago I wasn't sure humankind was ready to assume the responsibility required for its ultimate survival, especially in a dyadic universe. And I still don't know. To utilize the full power of volition requires a full measure of personal responsibility for the greater good. My mother surrendered her eyesight rather than see the world through what she thought were the eyes of evil. She didn't recognize that both healing and the ability to deny it dwelled within her all along, though she began to suspect this in later years.

Is belief in postmortem judgment and winnowing of the unjust really necessary to teach morality, or is moral behavior sufficient reward unto itself? Will moral behavior enhance individual and collective survival, or simply make the individual life a more pleasant experience to endure? The practices of all esoteric traditions urge their laities to rise above the ego level and discover contentment and morality in a purpose larger than Self. The fact that nature provided a means to achieve internal peace and nobility of thought for those seeking to do so is a significant evolutionary development. Nature doesn't respect the individual; the noble and lowly alike succumb to the shifting sands of time. Only the process is relevant to her; after all, nature is process. Therefore internal peace and nobility of thought must have survival value on the grander scale as well.

But how many are willing to sacrifice the external acquisition of power for the path of service and internal contentment? Such a path is supremely difficult to course, and precious few have both the courage and wherewithal to take it. For centuries the predatory and misguided in all esoteric traditions have in one way or another perverted the power of esoteric insight into the politics of guilt, sin, fear, and self-righteousness for personal gain. It's impossible to find one's transcendent Self within institutions whose purpose is to foster theocracy. Our species abandoned the ways of community consciousness and partnership with nature after the period of hunter-gatherers, and developed technologies and ego-based cultures for conquest and exploitation of nature, proclaiming ourselves in accord with divine will. Are we ready to again live in harmony with natural process, or must we fell every last tree and create a global wasteland before the error is clear? In a universe that learns by trial and error, human volition and the willingness to use it with enlightened benevolence for the greater good are key to ultimate survival.

Within the dyadic model, the source of all esoteric experience is resonance of the body with the zero-point field; that is to say, with nature itself. The more completely and coherently the mind/body is in resonance, the more expansive the perceived reality. Total resonance and coherence results in the nirvikalpa samadhi, wherein awareness is merged with the field and the sense of Self expands totally throughout the field. The experience is sufficient to alleviate all fear, and create a sense of joy and purpose—and the recognition that cruelty only harms the collective Self.

Initiates who experience the "rising of the kundalini" for the first time invariably report an extraordinary sensation. For a brief moment they sense they are about to die before bursting through to the explosively expanded awareness of the samadhi and the ecstasy that naturally follows. This is truly the dark night of the soul followed by the dawn. In confronting death, all fear is released, and kinship with the All-That-Is is established. I find it extraordinary that natural process allows for this experience. The most reasonable explanation for it is that the structure of life experiences that make up the ego-self, in a way, dies as transcendence is achieved. Then each cell of the body finds coherent resonance with the ground-state of being, and collectively reveals a greater truth.

The most celebrated mystics of all time, from Lao Tse and Gautama Buddha to Jesus and Mohammed, speak of the inner experience in markedly similar ways. The fact that their cosmologies were only metaphor doesn't detract from the power of the personal experience. Nor should the fact that, for centuries, followers have construed each metaphor literally, keep us from exploring the esoteric experience anew.

When you enter one of the great cathedrals of the world, what strikes you is the extraordinary ambiance. It is one of peace, tranquility, and awe; the atmosphere is overwhelmingly palpable. This, I would suggest, is the manifestation of ubiquitous resonance. This is a kind of communication with worshipers through the centuries who brought forth their highest aspirations, hopes, and beliefs in meditation and prayer, the edifice maintaining the resonance of the information they imparted with their thoughts. This is their spiritual legacy.

For years now, the mystical idea of connectedness and oneness has seemed to me inextricably linked to the concept of the universe as arising from an unlimited field of energy, without time, omnipresent, resonating with and reflecting each action in the manifest world. I also think of it as possessing the seeds of awareness and volition. Except for anthropic form and omniscience, the field is a reasonable approximation of the way our distant ancestors imagined their deity to be. In this way we have created the gods in our own image.

The significance of the zero-point field and its relationship to spiritual experience burst into my awareness, strangely enough, late one night in 1985, high above the plains of Texas. I was returning to Florida from a meeting in California, and outside the small, oval cabin window, the night was moonless and filled with stars. As I relaxed in the darkness, my mind lost in the heavens swaddling the plane, I was reminded of experiences years ago in *Kittyhawk*. A strange familiarity washed over me. I again experienced the same sense of connectedness, again felt a part of the processes around me. And the heavenly bodies possessed scintillating halos. Then it dawned on me: Stars don't have halos, nor do they twinkle in deep space or at 30,000 feet. I knew my mind was playing tricks. But at the same time, something suddenly made sense. As I sat there surrounded by the heavens, I experienced a kind of intellectual synthesis. It was as though I was creating a mental picture of this impalpable energy field that causes matter to correlate nonlocally, and brains to perceive information nonlocally. In this relaxed state, questions and answers seem to drift together into a new pattern.

Though the experience didn't alter my sensibilities as profoundly as the experience in *Kittyhawk* 14 years earlier, it was heady enough. The quiet mind is capable of synthesizing information in powerful ways. The mystical experience originates in perceptions that seemingly have no external, physical source. This is likely the result of coherence in the brain, quieting the noise so that the perception of nonlocal resonance reaches the threshold of conscious awareness. In the nirvikalpa samadhi, most of the brain and body is likely in a ground-state of resonance with the zero-point field. The result is pure awareness and loss of the sense of Self.

In the psychic bands of consciousness below the samadhi state, additional information is being received through resonance with other physical objects. As one moves from state to state, resonance moves from the ground state of brain/body awareness to successively more active (and differentiated) states where resonance correlates with other objects. In the existential and lower states of awareness, the subtle perceptions of resonance are swamped by the activity from normal sensory information and thought processes, but are nevertheless there, operating below the level of conscious awareness. Examples of this are the studies that show that twins frequently behave and dress in concert, though they live miles apart and are not consciously coordinating their activities. This is what was tumbling through my mind high over the Texas plains in the seat of that airliner.

An early, unpublished paper by Harold Puthoff on the subject and the more recent work in quantum holography by myself, Marcer, and Schempp are seminal in suggesting that resonance with and through the zero-point field as the source of all mystical experience and psychic functioning. For several years now laboratory work has demonstrated resonance between humans, between humans and matter, and between simple organisms, but with no cohesive theory to account for it. Resonance with this non-spacetime field, and the mechanism of the nonlocal quantum hologram, is the only theory today that can explain the phenomenon in light of the fact that signals carrying such information do not obey the inverse square laws of electromagnetic theory, nor can they be attenuated by any type of shielding, such as a Faraday cage. The theory is testable and of course can be falsified, but it does explain replicated experimental data that hasn't been explained otherwise.

Religious and mystical experience, and all psychic effects, are, by this way of thinking, a result of the individual bringing nonlocal information to the level of conscious awareness. When perceiving information, the individual's subjective experience and the *interpretation* of those experiences will conform to the beliefs that their cultural conditioning allows as valid and permissible. The left brain desires to interpret perceptions in a coherent, consistent, and logical way. Therefore, the nonlocal information is connected to information already in memory, so as to provide that logical, consistent, and socially acceptable thought structure wherever possible. When in these states and intending a particular action, resonances within the field are altered, and energy from the zero-point field available to create the desired effects.

We accept without question that thoughts expressing desires and intentions create action within our own bodies. Every time we lift a finger, or become thirsty and reach for a glass of water, the brain/body responds. The neural connections do their jobs, and intended action takes place. We are less willing to accept, however, that similar effects can be enacted outside the body—that is to say, nonlocally. The neural connections that make this possible are either lost or become inactive after childhood in most people, as they are not exercised and maintained. But modern studies clearly show they can be maintained and reactivated with proper training. People of all ages can become psychokinetically active with proper exercises and dedication. Some even do so spontaneously. Mozart used a different neural pattern than Einstein, and both were different than Picasso's. But I believe it is quite possible, with proper conditioning beginning prenatally, to activate and maintain neural connections and functions in the brain that will allow a particular individual to have the best qualities of a Galileo, Mozart, Einstein, Baryshnikov, Norbu Chen, and Uri Geller. It's just a matter of initiating such exercises before the brain begins to abandon and prune unused neural connections during childhood. And this is the key to our conscious control of evolution.

For more than a decade I've maintained an interest in accelerated learning developments, having confidence that a way to better utilize our latent capabilities would eventually emerge. In late 1993 I embarked on research concerning specific techniques and their results. Although research has been ongoing in the field of subliminal learning in different parts of the world for more than 35 years now, the work of educator Makoto Shichida in Japan is of particular significance. Today his results are being successfully replicated in the United States.

Using prenatal techniques similar to those pioneered by Shinichi Suzuki to precondition a fetus to musical sounds, Shichida used words as well to precondition for postnatal learning—that is to say, to store desired information. Promptly after a child's birth Shichida would employ simple game-like exercises with flashcards and sounds to teach numbers, words, and ideas. Other exercises strengthen both long-term and short-term memory, and pattern recognition. The results have shown that prodigious learning skills are available and can be maintained from the cradle. Eidetic imagery[2] is strengthened, and maintained so that the children are soon capable of reading with virtually perfect retention nearly as quickly as they can turn the pages of a book.

One of the most interesting (though not surprising) results from the point of view of a dyadic model is found in flashcard matching games. Shichida found in his studies that children quickly achieve almost perfect accuracy in guessing the identity of flashcards that lay face down.

Their intuition—that is to say, coherence between conscious and subconscious functions—quickly develops to the point that information from the subconscious, the rational functions, and nonlocal sources seems equally available. There seems every reason to believe that a wide range of disciplines or skills can be cultivated in this fashion. It is well documented that the rate at which learning takes place rapidly declines once children begin school and learn to read through rote memorization and vocalized reading—perhaps the most ineffective of all techniques. It is the simultaneous development of the rational and the intuitive, equal emphasis on right and left hemisphere functions, that most dramatically enhances the rate of learning. Although by the time of puberty an individual's thinking patterns and basic belief system have been established, there is ample research to support the belief that with proper training and practice, brain functions, undeveloped early in life (such as eidetic imagery), can be recovered.

Few individuals study their internal feeling sense carefully enough to be able to distinguish the particular sources of internally sensed information. This is true primarily because our traditional sources of authority in Western civilization, both ecclesiastic and scientific, dismiss internal sensing as unimportant. What we value is the sheer candlepower of the intellect. Thus, we're astounded when accomplished musicians, artists, writers, and poets, even genius mathematicians and scientists, perform as they do. Their powers seem supernatural, magical. But the source of their power is the intuitive process and creativity of the subconscious. Whether they use it consciously or not, this is the faculty they employ. And with practice, any individual can become more intuitive and creative—but first one must learn to distinguish precisely from where this internal sensing is coming.

Intuitive information has two sources. The first is the result of stoking one's brain with information, posing a question, then letting go of the issue (sleeping on it). The subconscious then attempts to resolve the issue and present a solution to conscious awareness, perhaps through an *aha* sort of insight, perhaps through the symbolism of dreams, or by producing a "synchronicity." One needs to be alert to capture the more subtle intuitive responses, but they are always there. This subconscious capability suggests that the brain learned how to do problem-solving before self-reflective awareness evolved, including how to use intentionality to create certain outcomes.

The second source of intuition is nonlocal resonance with another person or object. True creativity may be defined as creating something new for the very first time. But a startling "new" insight that the perceiver believes to be totally original, may indeed be nonlocal resonance with the true originator.[3] In light of this, it isn't surprising that scientists, writers, and musicians sometimes accuse one another of stealing original theories, story lines, or songs. We resonate with themes and ideas when our subconscious is seeking the answer to a question we have posed. This also explains the so-called hundredth monkey phenomenon, which suggests that our animal cousins and our more primitive ancestors function in a similar way.

An unfortunate trap for those unfamiliar with internal sensing is the inability to distinguish between memories stored in the subconscious and true intuitive problem-solving. They may have much the same feel. A vague discomfort when meeting someone new might be an intuition that the individual's intentions will not best serve your interests—or it might be just a Pavlovian response to the person's nose resembling the nose of your overbearing Aunt Minnie. Likewise, "love at first sight" could have more prosaic origins.

29

Today I live a very different life than I have in the past. There is an internal world I have come to utilize more fully through the years, a world in which I feel more in sync with natural processes. Yet I'm endlessly amazed when I discover a new level of intricacy in the environmental patterns. But the journeys to discover the worlds of the Shaman are essentially complete. The task now is to make better sense of it all as new experiments in science and probes of the cosmos provide new understanding of humankind and our place in the cosmos. For a year following the first publication of this account, I was content to sit on the patio with my entourage of Schnauzers, watching the sea breeze sway the summits of the pines that shroud my home. A small family of horses grazing beyond the perimeter of a pine fence provided quiet amusement with their interactions and social ordering. On weekends my son Adam might visit, stirring the brood of dogs into a frenzy. On other occasions I chose to visit with my older children and grandchildren as frequently as possible. They provide the meaning for years of exploration. After 20 years of city life, of balancing the demands of business, research, the institute, and family, my life took on a firmer ballast. Arcane thoughts are more easily explored in this country setting. But retirement is not part of my character, and soon the siren call of paradoxes and incomplete understanding reasserted itself.

Not unlike other young people his age, at 23, Adam is very adept at the computer. On one of my machines he often played solitaire, which essentially simulates the physical game without using actual cards. One day while watching him play, it occurred to me just why it isn't a perfect simulation, and why the differences were important to my understanding of consciousness.

In either game (the one played with real cards, or the computer simulation) you begin with eight cards face up and 44 hidden from view. Assume that after a few plays two red cards of equal value are visible above equal stacks of hidden cards. Only one of the red cards can be moved in order to turn over the card below it. Is it important which one of the red cards is chosen? One, both, or neither may have a playable card immediately beneath it. According to the mathematics of probability theory, it's immaterial which red card is moved, as a playable card is equally likely under either.

The computer simulation of solitaire uses pseudo-random probabilities to select the value of a particular card as it is turned face-up. The computer doesn't deal and stack the cards before play begins as in the real game, even though your computer display would have you believe it does. In the computer game, it truly *is* immaterial which red card is moved. All the remaining cards have *exactly* the same probability of being selected to be the next card revealed. In the game using physical, well-shuffled cards, a playable card may already exist under one or both of the red cards. It was positioned when the cards were first dealt. The question is therefore one of knowing what already exists.

In the absence of nonlocal intuition we cannot know with our five external sensors which card to select. With just the five senses, the computer game and the actual game are mathematically equivalent. But with resonance and *feeling*, one can often beat the odds and select the right card when one is more advantageous than the other. Laboratory subjects have been beating the odds at selecting cards since J.B. Rhine began studies in the 1930s. And long before then, gamblers in Las Vegas were well aware of this mysterious and elusive sense.

However, in the computer simulation, the cards don't actually exist, so intuition doesn't come into play. Yet individuals experienced in the Shichida training can select the preferred card in the actual card game almost every time. Thus the computer game isn't a true simulation of the physical game. Probability theory based upon random-number mathematics doesn't predict the correct answer when nonlocal intuition is present.

This is precisely the same problem we had with Schrödinger's cat. The cat really did *exist* in one state or another; which one was just not known. But the state was *knowable*, either by opening the box or by nonlocal resonance. The mathematics and the knowing were all in the mind; the state of existence was confused with the probability of being able to know the state. In computer-generated solitaire, knowing the next card is not possible because it doesn't yet exist. The process for selecting one is pseudo-random; therefore, intuition doesn't help in the computer version.

However, *intention* can. Robert Jahn at Princeton, physicist Helmut Schmidt, and Dean Radin, formerly at University of Nevada at Las Vegas, and now at the Institute of Noetic Sciences, have demonstrated with long-running experiments that intentionality can affect physical systems in general, and computers in particular. J.B. Rhine's early work has been confirmed. As their statistics show, intention can influence random processes in mechanical devices. The statistics are neither as startling nor as bizarre when compared with what Uri and Norbu can actually do, but they are of equal significance in our understanding the basis of reality.

Which leads us to the popular misconception among mystically inclined folk that precognition or fortune-telling is possible. Not in a dyadic universe. Even some of the early work in card-guessing was interpreted as precognitive guessing, which would be possible only if the cards were already ordered. In a dyadic model of reality, the same results would be interpreted as psychokinetic—as intending instead of knowing. Because time moves only forward and all life processes are nonlinear and include choice, the future is not fixed, and therefore not knowable. But it can be influenced or even created to a certain extent. Accurate prophecy is more often really self-fulfilled prophecy. What is knowable through nonlocal intuition and expanded awareness is an expanded sense of now, not a sense of the future. Processes in motion that have a certain amount of stability can be predicted a limited way into the future by an individual with expanded awareness. This expanded awareness of existing processes already in motion gives sensitive people the feeling of obtaining information from the future, a sense of premonition. But because volition is real and people do make choices, those choices affect the future of the process. And the ability to precognize the future diminishes rapidly with time unless very stable processes are involved. A particular future is at best only a probability until it actually happens.

In New Age culture and weekend self-help training, it's now popular to employ aphorisms such as "think only positive thoughts," "be careful what you pray for," "you create your own reality," "matter is just a dense thought," and the like. But do any of these notions have any validity? In my opinion, they're somewhere near the target. But they need to be more carefully examined.

Too much of our human thought is tantamount to snow on the television screen. Our brain creates energy pulses, which we experience as information. Thought is simply information coming to the screen of conscious awareness, and being aware of information doesn't, by itself, do anything but allow one to know what one is resonating with. It doesn't kill Schrödinger's cat. It does little good to attempt to suppress the negative and overlay it with sweetness and positive thinking, if troublesome thoughts keep surfacing. In this case we merely sublimate a problem that will likely surface under stress. We must accept responsibility for our thoughts, whatever they are; they are ours alone to manage. If we don't like them, or they aren't productive, we can and should change them. Meditation helps. But if we really can't, then professional help is in order.

But thorough and careful thinking pays off. The concept was ingrained in me from childhood through my career with NASA. It is really quite an amazing phenomenon that practicing patterns of conscious activity causes the subconscious to then habituate those patterns of thought. A major portion of my training in the space program had to do with "what if" scenarios. What if this goes wrong, or what if that component fails? These mental exercises were, in a sense, negative thinking. By contemplating in this manner, we could reveal what components of a system were likely to fail. This was a necessary intellectual process we had to engage in. But did they promote failure? Of course not.

This is no more negative than checking the weather to see if an umbrella is needed, and then checking the umbrella to see if it has a hole. We were simply becoming aware of dangerous situations and potential problems, then preparing to handle them should they occur. The *intention* was to create success and avoid failure. By intending to be prepared, and then following through, virtually impossible situations were salvaged by concerted planning and action. It is the intention behind action that's important; the rest is just mechanics. The systems we were most concerned about seldom failed—it was those we were complacent with that caused problems.

A popular misconception holds that merely contemplating potential failure begets failure. Of course, this too is false. Though failure-mode analysis has been successfully applied to organizational problems as well as mechanical problems, commercial applications are difficult to market due to this bias. Only in Idealist models could such casual mental activity have this effect. It is true, however, that if one is stuck in negativity, viewing every situation in light of why it cannot possibly succeed, then one likely cannot succeed under these circumstances. By reinforcing ideas and giving them additional energy, one is impelled in the direction of the idea. Your worldview is precisely defined by the ideas and memories contained in the subconscious, which directs the course of conscious thinking. Phobias are excellent examples. When dwelling in the ego level and below, fear is magnified and one is propelled toward situations where the things we fear most are present at every turn. One best learns to overcome irrational fear by facing it directly, discovering that it is but a shadow that vanishes in the light of understanding.

It's quite possible to take the positive-thinking idea to extremes and to float through life in a bucolic haze. Life has its trauma and heartache, its pains and sorrows, and nothing is gained by glossing over unpleasantness. The most successful formula for dealing with the negative aspects of living comes from the skilled mystic who practices emotional detachment from the vicissitudes of life, maintaining amused vigilance over both success and failure in equal measure. By gaining such control, they gain control of their lives. They understand we are all engaged in the seemingly eternal cosmic game of creating a universe through trial and error, and learning from mistakes or unwanted outcomes.

Toward the Future

30

For years now I've made my living on the lecture circuit and as a consultant. All the while I've assimilated the results of ongoing research in consciousness studies into my work whenever appropriate, generally to assist individuals in finding a broader viewpoint for themselves. As a result I am regularly asked about my views on subjects as diverse and far out as whether we were followed by UFOs on the lunar voyage, to the nature of guardian angels. Because I try to take the attitude that there is no such thing as a stupid question, only stupid answers, I attempt to address each question seriously. By demonstrating the courage to ask unconventional questions, people indicate an authentic desire to find answers, albeit sometimes in strange places.

In the late 1980s I observed a marked change in the general attitudes of people and the questions they asked. Consciousness was becoming a topic of serious concern to scientists. Professional businesspeople who had shown little interest in such arcane subjects were suddenly asking questions too. It seemed as though worldwide there had been arising a deep sense of malaise, as people began expressing their concerns, oftentimes simple intuitive ones about the future course of civilization. They were puzzled and felt that traditional answers were no longer adequate. Judging from my personal observations, this was, and still is, a global phenomenon.

A few of us have been talking about paradigm shifts for more than 30 years, and it seems we now have one emerging. Suddenly there is renewed interest in spiritual matters, family values, and scientific explanations for the mystical experience. We want to know how to improve this world we were born into; we want to know how to avert what is perceived by many as the approach of man-made apocalypse. We also want more order in our lives. The primitive fight-or-flight response to threat and conflict has only been modified in small measure through the millennia to provide reasoned, negotiated outcomes for human society. Today's cultural cosmologies and value systems that are incompatible with an evolving universe have to yield as science finds ever-new confirmations of that theory.[1] But the religions from which our personal values are traditionally derived have sought to remain changeless.

With populations doubling in each quarter-century, and the subsequent strain on planetary resources for an acceptable level of comfort for this exploding population, we find ourselves in trouble. And with most of the world seeking the lifestyles and affluence of the industrialized countries, global civilization itself is in a dilemma. It is, I believe, the monumental size of this problem that is resonating as concern at the intuitive level.

Many predicted that the end of the millennium would bring about an apocalypse in the vein of the Second Coming. Others believed that it would initiate intervention by alien intelligence, and still others anticipated sage solutions by guardian angels and channeled entities. Some believe our dilemmas are merely political. By whatever process individuals have arrived at their conclusions, there is general consensus that the problems are real. Through the years I've come to the realization that the problems of postindustrial civilization are not only real, but that they are also severe, and deepening.

However, such problems are of our own making and can only be solved by human beings using their individual and collective creative resources in more constructive ways. We possess ample resources for the task. But first a readjustment in our thinking is in order; a worldwide change in the interests of creating a sustainable civilization, not just for ourselves as individuals, but in the larger sense.

Guardian angels, channeled wisdom, and divine revelation are traditional explanations for the perception of the deep cultural resonances and collective consciousness available nonlocally to any individual at any time. Of course, such information will be interpreted and assigned meaning according to the beliefs and biases of the percipients. The more rich and more varied the information base, and the more closely that belief is aligned with natural process, the wiser and more meaningful the interpretation can be. I am both delighted and amused that channeled wisdom seems to have become more earthy and practical during the 40 years I've observed such phenomena. Ageless wisdom based upon integrity, tolerance, and goodness is still pertinent to the modern experience.

I was often asked my opinion as to whether or not the end of the millennium would bring about any extraordinary event, or whether it innately carried any special meaning. There seems to be suspicion or hope that some sort of extraordinary intervention will magically relieve us of our problems here on Earth. Many assumed that some deep significance lay in the date itself. When asked whether I believed it did or not, I generally answered the question rather curtly: only if you lend it meaning. Two thousand years is just an arbitrary number on a man-made calendar. Again, nature knows nothing of time, only process. That date has now passed, and the same problems still need to be solved.

The transition from traditional reliance on external authority to save the day, be it God or government, is a difficult one. The ego-self is a master at avoiding responsibility and looking to others for satisfaction, often evoking the basic fight-or-flight impulse. Similarly, the ego is skillful in avoiding recognition of our interconnectedness, preferring to be concerned only with satisfaction of Self. But such impulses are inappropriate in today's crowded world. Guardian angels, channeled information, alien visitation, or mere prolonged government subsidy are all forms of postponing the inevitable day when we must own up to our responsibilities for Self and the full use of our inner resources for the larger good.

Traditionally we've been taught to respect authority. Yet "authority" itself is in a dilemma, for traditional institutions are not structured to handle these modern issues. So it's not surprising that today we see respect, not only for authority, but also for civilized behavior itself breaking down. Conditions are accurately interpreted as heralding a paradigm shift, but it also has the characteristics of a bifurcation point that is unpredictable as to outcome. The crisis is of both existence and knowing—too much existence and not enough knowing. The antidote lies in knowledge, awareness, finding meaning and purpose for one's existence that is beyond Self, and then exercising the personal responsibility to carry out that purpose. It seems clear that continuing to seek only material ends, to glorify unlimited economic growth in the face of the hazards that such a worldview entails, is foolhardy. I have full confidence that as a species we can transcend those limitations and create a sustainable society, though the challenge isn't an easy one.

Our species seems capable of evolving the attributes the ancients ascribed to deities. *God sleeps in the minerals . . . and thinks in Man.* But is humankind actually ready to accept the responsibility for this evolutionary leap? Are we ready to assume godlike status? It surely lies in the offing, provided we create a nurturing environment wherein we can understand and live harmoniously with natural processes, and develop all the capabilities already resident within us. But we have yet to grow into the brain nature provided.

Because we truly do have volition, we can also, through foolishness or ignorance of natural processes, terminate our existence. Within my lifetime this has become not only possible, but more than minutely probable. We live in a trial-and-error learning universe; to intelligent beings, what does not work is as valuable a lesson as what does. Likewise, the growth and evolution of the universe toward wise, self-reflective awareness won't end, should *Homo sapiens* decide to behave like lemmings.

Barring nuclear or biological apocalypse, climate change, and/or total economic collapse during the 21st century, where does the future lead? What is our destiny as a civilization? There are certainties that lie in our distant future. Eventually we must leave Earth—at least a certain number of our progeny must, as our sun approaches the end of its solar lifecycle. But just as terrestrial explorers have always led the way for settlers, this will also happen extraterrestrially. Earth is our cradle, not our final destiny. The universe itself is our larger home. And without doubt we will meet other species along the way—if we haven't already—who, like ourselves, have begun to explore the cosmos.[2]

Will they be hostile to our kind? Probably not. By the time a species is capable of exploring the deep cosmos, they have surely learned more about themselves than we, and the value of benevolence. It's safe to say they will be intelligent creatures who already fully recognize that all life is connected, though it may have different form and appearance. They should already recognize that the only way to conquer the universe is to know its ways and align oneself accordingly.

However, the future isn't knowable, prophets and soothsayers notwithstanding. We will not discover our future; we will create it. In the

Physics of Immortality, Tipler speculates that one day humankind will expand its presence from the earth to the solar system, from the solar system to the Milky Way galaxy, and from the Milky Way throughout the universe at large. In the process we will gain greater utilization of the energy resources available at each stage. And he even speculates that we might be able to utilize energy and knowledge to prevent the big crunch, which some cosmologists believe heralds the end of this universe. He also speculates, as I mentioned earlier, that the objective information from each being, which still reverberates throughout space-time, could be reassembled and programmed into a giant machine, thereby emulating each individual that ever lived. From within the machine one could not, in principle, tell the difference between the original Self and the subsequent emulation. Physical immortality would thereby be achieved: Anyone who has ever lived would one day "live" again.

But I have doubts about such a scenario. I agree with Tipler's assessment that the destiny of humankind is to explore the universe, and perhaps even to manage its fundamental processes. But his argument for immortality is based upon capturing "objective" information about the individual. He uses the strong artificial intelligence argument that knowing is just computability, and objective information the only valid information. But I believe that information must be perceived subjectively, and the only valid meanings come from our conscious Self. We may not have immortal bodies, but we are eternal beings anyway, because the information from our having passed this way already resonates throughout the universe. We don't need a computer to capture it—I know of people who are already doing so.

The course of my life has been largely directed by a tendency to pursue ineluctable impulses that lead to vocations and avocations I was good at. All my life I had been both familiar with and comfortable in the mechanical world of engineering and science. Though religion wasn't something I was drawn to viscerally, I've been curious about it all my life, as it was a significant part of my upbringing. That is to say, I could study it in somewhat the same secular manner I could the dynamics of an airplane or the model of an atom. By the same token, my scientific pursuits have been colored by a tendency to see phenomena in the context of the Big

Picture, perhaps with an unseen hand quietly influencing the machine. It's as though I was inevitably pulled to see the relevance and connections between seemingly disparate worlds, to be bilingual in the language of physics and the word of the scriptures. Quantum physics is the language the Occidental scientist uses to find the ghost in the machine of the everyday world. Without it he cannot explain it so fully. Quantum physics is the language one uses to express the mystery and miracle of simple sunlight.

All mysteries are man-made, a product of our desire to know, and in the larger sense, the desire of nature to know itself. And all mysteries are, therefore, left for man to be solved as well. As dualism and paradoxes appear in our thinking, it becomes necessary to expand the old language, or to create a new one, and rise to a new level of perspective. Paradox is not resolved by the same level of thinking that created it—such is the condition of both the world of quantum physics and the mystical experience. Thesis, antithesis, synthesis. Sometimes more is required. This is where the physicist seems to require the language of the mystic, and the mystic the findings of science. Religious and scientific journeys are both really inner explorations.

I believe humans undertake most major journeys for reasons that are buried in the subconscious stream of life—the ineluctable impulse. Often we leave home in the interest of finding answers, and in the process we create new questions. This is why we explore space: the better to see ourselves in the vast scheme of a vast universe. To explore is to expand the horizon. Through this process we have found more questions to pose than we are presently capable of answering. But in this grand effort, we have fallen in closer step with the Tao. To send explorers into space is to participate in the ancient human drama, furthering the sphere of our human perceptions and intention. The dream we secretly dream is to reveal our own significance, to make the universe in our own image, just as we made the gods. But we must go there first.

These modern-day journeys, no less than their ancient predecessors, are the actualization of the mystical journey. Though the Hubble space telescope has helped to more clearly reveal an incomprehensibly large and continually expanding universe, we humans insist on our own divine importance. Centuries ago we discovered that we are not the geographical center of our tiny solar system, much less the cosmos; now it's quite certain that neither are we the biological center. In the face of this we try to find significance for ourselves. With awareness that the individual lifespan may be but a brief crack of light buttressed by infinite darkness on either side, we insist that the individual human life has some purpose in the larger scheme of things. And indeed it does, if we use our limited time to make it so.

Back in the 1960s there was, for a time, reason for me to hope that I could be one of those astronauts who would make the next logical step in humankind's saga to explore the cosmos. My doctoral dissertation proposed a navigation plan for such a mission. And there was reason to hope that our government would find this a worthy priority after our nation's gallant series of missions to the moon. The mandate set forth by President Kennedy in 1961 came in both on time and on budget—a feat almost unheard of for the federal government today. In the aftermath of such success, I also felt it wasn't unreasonable to hope I could be part of a mission to Mars. Plans for the space shuttle were beyond the drawing board, as mockups and tests were already being conducted. Soon, it was hoped by many at NASA, there would be a space station, followed by what would evolve into the next giant leap for humankind.

But there would be no manned Mars mission during my career.

The remnants of our national effort of the 1960s and '70s is but a tiny fleet of space shuttles that have been criticized for being neither inexpensive nor especially versatile, a few unmanned missions that have sent back pictures and data from our neighbors in the solar system, and an international space station still under construction. Many scientists have been vociferous in their criticism of the shuttle program, claiming that manned missions in near space have amounted to little more than clowning around in zero gravity. This is absurd, as commercial space on shuttle missions is highly sought-after. The work that has been done has been mainly in fields of communications, basic science, and medical research, none of which are so easily promoted as something as visual and bold as a lunar mission, or pictures from a relatively inexpensive unmanned probe flying by a colorful distant planet.

This is not to say that the hiatus in manned space exploration has amounted to lost opportunity. The solar system will change little by the time we are ready to make our next foray. Perhaps history will show that this time during which we've necessarily turned our attention to the problems of our planet will prove useful in garnering a consensus of will and a new cooperative effort, allowing us to pool resources with other nations, making it truly an indiscriminate journey of Earthlings, not separate nations.

When we one day venture to the red planet and the cameras are turned back on that shimmering blue dot, the explorers on that spacecraft will neither be willing nor capable, I believe, of referring to themselves as either astronauts or cosmonauts, or claim any nationality with heartfelt patriotic pride. Their nationalism will dissolve as Earth recedes from view, and man-made boundaries fade with distance. They will only be able to call themselves explorers from planet Earth.

But until that time comes, until the day presidents and prime ministers around the globe propose a mandate stating that humankind will make that next giant leap, we will dream and continue to make modest forays into the near heavens. A more ambitious day will inevitably come. Perhaps it will be our children or grandchildren, or even their children, but one day a craft from this shimmering blue dot will lower into a pale red Martian horizon. As the engines blow away dust, and our progeny prepare themselves for the gravity of another world, they will slowly open the door of their spacecraft and walk beneath the same sun that rose that morning over New York, Moscow, Beijing, and New Delhi. They will do this so that their children's children may one day live and work there.

Months or years later, they will return home to a grateful planet where they will write about what it was like walking about on the face of that world. The work will be translated into the various languages of the world. More men and women will follow, and experiences and stories will accumulate in their wake. All the while, new technologies will be born, and the internal and external frontiers will be pushed back a little further. Then, gradually, imperceptibly, but inevitably, the shimmering blue dot will slowly recede in the view of the spacecraft that will carry our children's children throughout the ghostly white of the Milky Way. Still others will follow, and with them the ancient stories of their predecessors. Then they will leave the galaxy in order to make themselves in the image of God.

But this is a future that has only been imagined. It must now be created. From their distant perspective, the immensity, the beauty, the intricacy, and the intelligence of the cosmos will be palpable to our progeny just as it was to me. All I can suggest to the mystic and the theologian is that our gods have been too small; they fill the universe. And to the scientist all I can say is that the gods do exist; they are the eternal, connected, and aware Self experienced by all intelligent beings.

Epilogue

Revision 4, 2007

It has been a decade since this book was first published, and two years since the third edition. The rush of civilization toward a bifurcation point, in the vernacular of systems theory, or a tipping point in more common language, has not abated. A major problematic event in our unsustainable growth is the specter of *peak oil*. Peak oil is that point in time at which the discoveries of new petroleum reserves each year are less than world consumption. That event is now projected as imminent by many petroleum geologists. In 1972, during a speech at an International Rotary Convention in Lausanne, Switzerland, shortly after my epiphany in space, I raised concern about the finite supply of fossil fuels, and its role in sustainability. That reality will be upon us in short order. It is my hope that it will cause citizens of the world to awaken to the larger issues. We are overwhelming our planet's nonrenewable resources with our numbers and consumption patterns.

As this account is written, Adam, the youngest of my brood, has finished college at a prestigious film school and embarking on a career in the film industry. Although his mother, Sheilah, and I are no longer married, they have together produced from historical recordings a series of CDs and a DVD that captures the essence of my insights and concerns from the lunar voyage. In working together to tell that story using modern recording technologies, we have tried to practice the principles and values embodied in service to the greater good. As the 21st century begins to unfold, I find myself greatly concerned with the significant increase in violence and lawlessness in the world, which detracts from the critical need to collectively address the protection of our environment, and the larger issue of the sustainability of our civilization. Actually, it seems, we humans are not as civilized as we would like to think. The world population more than doubled in the 20th century to more than 6 billion people, and it is consuming Earth's natural resources at an outrageous rate. By some very reasonable estimates, our global natural resource base cannot support a population of more than 2 billion people consuming at the level of Western cultures. Clearly, major changes in thinking and lifestyles are required as developing nations naturally aspire to the consumption of the West. Yet our political system is now punishing dissent, freedom of inquiry, and efforts at preservation—tactics certain to slow the drive toward sustainability.

My extended family, now dispersed across several of the United States, is strongly united in addressing these vital issues, by working toward simplicity, peace, and cooperation. They have provided magnificent encouragement and resourcefulness in helping further these goals.

The issue of sustainability, simply put, is illustrated by charting several of any measures of human activity. Beginning noticeably in the 20th century, *all* measures display exponential growth. It is patently obvious that exponential growth cannot continue indefinitely in a finite space. Something, and likely many things, must be changed. Obviously, population growth is the engine driving the system, and addiction to the Western consumption model is not far behind as the major issue.

In an important new book, *Plan B 3.0*, Lester Brown discusses the fundamental areas of consumption and behavior that we as a species must address and modify in order to assure our descendants a reasonably satisfactory lifestyle. The days when we can ignore the larger view of what our individual decisions and lifestyles are doing to the collective have now passed. Personal responsibility for the greater good must become the mark of an informed and conscious people, then instilled rapidly into those yet unaware. *Raising consciousness* has been a motto of the new age through several decades of civil rights, feminine rights, and minority rights activism. Though that phrase sounds trite, even passé, to many ears, the concept of setting aside prejudice and bias in order to support a greater good and a larger view is vital to our collective well-being in the near future.

For myself, retirement is no longer a part of my personal vision for the future. These issues that can be accommodated under the banner of *sustainability* are too critical and pressing to be ignored. The problem is not simple, but includes many of the deeply embedded concepts that constitute modern lifestyles, cultural traditions, and thinking. Untrammeled consumption and growth in all areas of life must be reexamined and subjected to critical thought. Material goods are certainly necessary for sustaining life, but just where does success turn into excess? Those are the fundamental ideas we humans must now address. The scientific and technological genius exhibited beginning in the 20th century has cracked open the lid of the proverbial Pandora's Box, and loosed upon the world the very seeds of our destruction. We must act rapidly to bring our human penchant for viewing material abundance as a panacea for happiness, under control. The greatest philosophic and religious teaching of all time has been ignored and perverted with regard to this issue. May our descendants forgive us these errors as we struggle to bring our Earth back into balance.

Postscript

It's Time to End
the UFO Coverup

When I went to the moon aboard *Apollo 14* in January 1971, I had an experience that changed my understanding of the cosmos, and it happened totally outside the parameters of NASA's astronaut training. During the return voyage, as I gazed at Earth during a rest period, I had an epiphany, a spiritual insight. The connectedness of all things in the universe became apparent to me at a level of awareness beyond visual perception and outside my scientific worldview.

Implicit in that revelation was the idea that the universe is more than just inert matter. There are forms of life beyond Earth, some probably more evolved than us.

After I retired from the Navy in 1972, I formed the Institute of Noetic Sciences to study human consciousness and ways of knowing, including the intuitive mode I had experienced in space. (Noetics means the study of consciousness.) I placed the idea of extraterrestrial life forms in the back of my mind, so to speak, as I went ahead with the mission of IONS.

However, the idea was reawakened in me forcefully through circumstances that arose from a trip to the scene of my childhood. When I was a boy, my family lived in Roswell, New Mexico, the site of a July 1947 event that later became known as "the Roswell incident." That incident involved the alleged crash of a UFO, possibly two, and their retrieval by the Army Air Force, along with their occupants.

The Army Air Force itself was the first to make public disclosure about it, but a day later it denied the event happened and attributed it to a crashed weather balloon. I'd heard rumors around town of the UFO crash, but solid evidence for it was lacking to the public.

Some years after my Apollo space flight, I went back to Roswell on a nostalgic visit. I saw a lot of friends, including some of the old-timers in the area who had been involved in the Roswell events in 1947 and subsequent to the crash. One person had been in the sheriff's department and had gone to the crash site to supervise vehicular traffic. Another had been an officer at Roswell Air Force Base, where the UFO debris and dead aliens were taken. These and other people had been silenced by military authority and told not to talk about their experience, with a threat of dire consequences if they did. By the time I met them many years later, they were ready to break their silence. They didn't want to go to the grave with their story suppressed and untold. They wanted to tell someone reliable, someone trustworthy.

Being a local boy and having been to the moon made me such a person in their view. They considered me reliable and trustworthy enough to whisper in my ear, so to speak, personal stories of their association with the Roswell incident. One of them was Jesse Marcel, Jr., the son of Major Jesse Marcel, the first military officer on the crash site. He had brought home some pieces of the crashed UFO and showed them to his family, including young Jesse (who has now written a book about it). Another informant was a member of the family of the funeral director who provided coffins for the alien victims.

I became intrigued, and then avid, about the possibility that Earth has been visited by ETs—space travelers whose activity long preceded and greatly exceeded my own. I continued to meet with UFO researchers and people who claimed to have experienced UFO sightings and even alien contacts. I was pleased to learn that even some of my fellow astronauts have had UFO sightings, such as Gordon Cooper, who has described his in print and on television.

Today, I can say on the basis of my reason and research that we are not alone in the universe. That is not a startling statement considering what science has learned through astrophysics and astrobiology. We know that the building blocks of life are found throughout the universe and it is therefore highly probable that other life forms exist. Some of them may be more highly evolved than us.

Astronomers have now identified more than 1,200 stars with planets orbiting them. Some of those planets appear to have conditions favorable for life as we know it here on Earth. Not only are other life forms likely, there may be a community of planets with intelligent beings more advanced than we are.

But have they actually visited Earth? That is where mainstream science draws the line of denial, saying there is no evidence of visitation. However, those of us who follow where the evidence of ufology—the study of UFOs—leads see a very strong case for ET contact, perhaps extending over millennia, as the so-called ancient astronaut hypothesis asserts. The ancient astronaut theme is found in myths and legends from cultures and societies around the world. Were ETs perceived by our pre-scientific ancestors as gods from outer space because they lacked our understanding of science and technology? Did our prescientific ancestors mistake ETs for deities and UFOs for "chariots" of the gods?

That aspect of the ET hypothesis is more conjectural than the modern evidence, which is persuasive to reasonable minds. Tens of thousands of visual and radar sightings by civilians and military people all over the world confirm that something alien is in our skies. Many of the witnesses are trained commercial or military pilots. Photographs, film, and video footage likewise confirm that.

Even more to the point is this: I have had additional confirmation from very high-ranking members of several governments that UFOs are real and that ETs have made contact with Earth. I am not at liberty to disclose the names of my informants, but they are from European nations, including the UK. Their sources of knowledge: their own militaries' studies of the subject.

I, personally, have had no UFO experience—not when I was growing up in Roswell, not as a pilot nor as an astronaut in space. Nevertheless, I am persuaded, utterly and completely, that we are being visited by extraterrestrials.

So why does the coverup continue? I suggest that it is time for disclosure by the US of what really happened at Roswell and other rumored UFO crash sites. President Obama campaigned on a promise of greater freedom of information and transparency of government. The question of ET contact would be a good place to keep that promise.

Afterword

As I write this afterword, the Institute of Noetic Sciences (IONS) is approaching its fiftieth anniversary. Over that half century, the Institute has vigorously pursued Edgar Mitchell's vision that the tools and techniques of science could help us gain a better understanding of consciousness and its capabilities. We have blazed new trails into the frontiers of consciousness via controlled laboratory experiments on psychic phenomena and by exploring the nature of transformative noetic experiences, the roles of gratefulness and forgiveness in healing, the efficacy of energy medicine, and many other initiatives that have expanded our understanding of the farthest potentials of the human mind, as well as pragmatic applications of those potentials.

One way to assess how our work has influenced the world at large is by recalling that when IONS was founded in 1973, concepts like the mind-body relationship, the role of spirituality in healing, and the mental and physical benefits of meditation were considered laughable nonsense by the mainstream. IONS provided some of the first grants for researchers to study these topics, and today not only are they taken seriously, they have become solidly mainstream as evidenced by health insurance companies reimbursing programs that teach meditation and by an increasing number of medical schools that include coursework on the role of spiritualty in health and healing.

When it comes to advancements in understanding psychic phenomena, these effects are still challenging to the status quo in some quarters, but the field has significantly advanced since the 1970s. Evidence of this can be seen in a 2018 article published in *American Psychologist*, the flagship journal of the American Psychological Association. The author, psychologist Etzel Cardeña of Lund University, presented the results of meta-analyses of over a thousand published psychic (psi) experiments in eleven categories. He concluded:

The evidence provides cumulative support for the reality of psi, which cannot be readily explained away by the quality of the studies, fraud, selective reporting, experimental or analytical incompetence, or other frequent criticisms. The evidence for psi is comparable to that for established phenomena in psychology and other disciplines. (Cardeña 2018,people have reported events that seem to violate the common sense view of space and time. Some psychologists have been at the forefront of investigating these phenomena with sophisticated research protocols and theory, while others have devoted much of their careers to criticizing the field. Both stances can be explained by psychologists' expertise on relevant processes such as perception, memory, belief, and conscious and nonconscious processes. This article clarifies the domain of psi, summarizes recent theories from physics and psychology that present psi phenomena as at least plausible, and then provides an overview of recent/updated meta-analyses. The evidence provides cumulative support for the reality of psi, which cannot be readily explained away by the quality of the studies, fraud, selective reporting, experimental or analytical incompetence, or other frequent criticisms. The evidence for psi is comparable to that for established phenomena in psychology and other disciplines, although there is no consensual understanding of them. The article concludes with recommendations for further progress in the field including the use of project and data repositories, conducting multidisciplinary studies with enough power, developing further nonconscious measures of psi and falsifiable theories, analyzing the characteristics of successful sessions and participants, improving the ecological validity of studies, testing how to increase effect sizes, recruiting more researchers at least open to the possibility of psi, and situating psi phenomena within larger domains such as the study of consciousness. (PsycINFO Database Record (c 1)

The same conclusion was reached in 2016 by University of California at Irvine statistician Jessica Utts, who at the time was president of the American Statistical Association. She wrote:

> For many years I have worked with researchers doing very careful work in [parapsychology], including a year that I spent full-time working on a classified project for the United States government, to see if we could use these abilities for intelligence gathering during the Cold War. . . . At the end of that project I wrote a report for Congress, stating what I still think is true. The data in support of precognition and possibly other related phenomena are quite strong statistically and would be widely accepted if it pertained to something more mundane. Yet, most scientists reject the possible reality of these abilities without ever looking at data! (Utts 2016, 1379)

Utts's last sentence may seem like an exaggeration, but the reality is that leading skeptics today insist that "there is no good reason to consider the data" (Reber and Alcock 2019, 8). In other words, the best argument that critics of psi research can offer today is to blithely ignore data. In a paper published in response to Cardeña's article, the same skeptics argued:

> If the physicalist-materialist framework of modern science is correct within the bounds of demonstrability and theoretical coherency—and everything that has been learned through science says that it is—the fact that claimed parapsychological phenomena are so grossly inconsistent with that framework suggests that they are all but impossible and that the claims made by proponents cannot be true. (Reber and Alcock 2019, 2)

That "cannot be true" assertion might have been justified if one were limited to a 17th-century grasp of physical reality. But we are in the 21st century, and we now know, as philosopher Andrew Westcome wrote in a response to the assertion that psi cannot be true, that *"every major paradigm shift in science was literally a violation of the basic scientific principles of the time"* (Westcombe 2019, 619; emphasis in the original).

The state of physics in the third decade of the 21st century does not yet adequately explain psi, but the picture of reality painted by modern physics is remarkably compatible with psi. Both domains involve connections that transcend the everyday constraints of space and time, and both domains also show that the act of observation influences aspects of the physical world. Some claim that these similarities are mere coincidences, but that argument seems implausible given the evidence for quantum effects observed in a wide range of biological systems (Ball 2011). It seems increasingly likely that the brain also exploits the more powerful information-processing capabilities of quantum phenomena (as compared to classical physical effects), in which case a clear roadmap is emerging that provides a way to understand psi in scientific terms.

In contemplating psi phenomena, Edgar proposed what he called a *dyadic model* of mind and matter. In philosophical terms, this model is perhaps better known as dual-aspect monism. It proposes that mind and matter are fundamental components of the fabric of reality, akin to two sides of the same coin. A long list of prominent philosophers have advocated dual-aspect monism (or a slightly different version called neutral monism), including Baruch Spinoza, William James, and Bertrand Russell (Stubenberg 2018). It has also been taken seriously in science, including by physicists Wolfgang Pauli, David Bohm, and Bernard d'Espagnat (Bohm 1986; d'Espagnat 1979; Heisenberg 1959).

Of greater importance, there is a mounting trend within science and scholarship today that is questioning whether materialism (i.e., everything is made of matter, including your mind) is sufficient to comprehend the fabric of reality, which must include the nonphysical phenomenon of consciousness (i.e., awareness, or first-person subjectivity). David Chalmers (2020, 353), a leading figure in the philosophy of mind, refers to this trend via a quip: "One starts as a materialist, then one becomes a dualist, then a panpsychist, and one ends up as an idealist." He explains:

> First, one is impressed by the successes of science, endorsing materialism about everything and so [also] about the mind. Second, one is moved by problem of consciousness to see a gap between physics and consciousness, thereby endorsing dualism, where both matter and consciousness are fundamental. Third, one is moved by the inscrutability of matter to realize that science reveals at most the structure of matter and not its underlying nature, and to speculate that this nature may involve consciousness, thereby endorsing panpsychism. Fourth, one comes to think that there is little reason to believe in anything beyond consciousness and that the physical world is wholly constituted by consciousness, thereby endorsing idealism. (Chalmers 2020, 353)

Edgar was initially drawn to explore the frontiers of outer space, which he did as one of only a dozen people to walk on the moon (as of 2022). Based on his mystical experiences and his astounding psychic experiences upon returning to the Earth from the moon, as described in this book, he saw a new vista that presented challenges even greater than that of the moon—the frontiers of inner space. As the epitome of an explorer, Edgar was always ahead of the pack and was regularly accomplishing "impossible" things while doubters remained stuck in old assumptions. As we can see ever more clearly today, his conviction that the exploration of consciousness would lead to new ways of understanding our role in reality was also far ahead of the pack. At IONS, we continue to look over the horizon by exploring inner space with the same spirit and enthusiasm that was the guiding light throughout Edgar's lifetime.

—Dean Radin, MS PhD, Chief Scientist, Institute of Noetic Sciences
February 28, 2022

References

Ball, P. (2011). Physics of life: The dawn of quantum biology. *Nature* 474, 272–274.

Bohm, D. J. (1986). A new theory of the relationship of mind and matter. *Journal of the American Society for Psychical Research* 80(2), 113–135.

Cardeña, E. (2018). The experimental evidence for parapsychological phenomena: A review. *American Psychologist* 73(5), 663–677, https://doi.org/10.1037/amp0000236.

Chalmers, D. (2020). Idealism and the mind-body problem. In *The Routledge Handbook of Panpsychism* (353–373). Routledge.

d'Espagnat, B. (1979). The quantum theory and reality. *Scientific American*, 158–180.

Heisenberg, W. (1959). Wolfgang Pauli's philosophical views. *Zeitschrift für Parapsychologie und Grenzgebiete der Psychologie* 3(2/3), 120-128. [Originally published in *Die Naturwissenschaf-ten*, 1959, 46 (24), 661-663].

Reber, A. S., and J. E. Alcock (2019). Searching for the impossible: Parapsychology's elusive quest. *American Psychologist,* https://doi.org/10.1037/amp0000486.

Stubenberg, L. (2018). Neutral monism. In E. N. Zalta (ed.), *The Stanford Encyclopedia of Philosophy* (Fall 2018). Metaphysics Research Lab, Stanford University. https://plato.stanford.edu.

Utts, J. (2016). Appreciating statistics. *Journal of the American Statistical Association* 111(516), 1373–1380, https://doi.org/10.1080/01621459.2016.1250592.

Westcombe, A. (2019). I do not think that word means what you think it means: A response to Alcock and Reber's "Searching for the Impossible: Parapsychology's Elusive Quest." *Journal of Scientific Exploration* 33(4), 617–622.

Notes

Chapter 3

1. One of them, Lieutenant Richard Truly, later Admiral Richard Truly, would one day become a NASA administrator.

Chapter 6

1. Kissinger was so impressed that during the post-flight White House meeting with the president and Mrs. Nixon, he invited Karlyn and Elizabeth into his office for a chat. Fortunately, it would seem, we didn't go to Hollywood for a studio tour with Mr. Douglas.

Chapter 8

1. We would later learn from photos that we were in fact but a few feet from the rim, a walk of perhaps five more seconds. The undulations of the terrain, however, prevented us from seeing where we were.

Chapter 9

1. Ours was the last lunar mission to be quarantined, as the moon was confirmed to be a thoroughly dead planet.

2. The results of the experiment were published in the June 1971 issue of *The Journal of Parapsychology*.

Chapter 10

1. Al Worden was a command module pilot.

2. Recent evidence regarding missing mass in the universe (dark matter), increasing acceleration of the expansion of the universe (dark energy), violation equality of charge and parity numbers of particles

and anti-particles resulting from the big bang, plus a reinterpretation of the red shift, are leading some scientists to conclude that the big bang theory may be incorrect, because it cannot account for these phenomena.

Chapter 11

1. In 1993 the Institute of Noetic Science published his work, *Spontaneous Remission: An Annotated Bibliography*, written with Caryle Hirshberg.

Chapter 12

1. The linguist Noam Chomsky believes the origin of language is just less than 250,000 years old.

Chapter 13

1. Curiously enough, my mother never underwent corrective surgery, though throughout the years her eyesight did slowly improve so that she wore a prescription with less correction just prior to her death than before she met Chen.

2. I've come to the realization that one must first cultivate a composed and serene internal state and take responsibility for one's own wellbeing. Only then can such external assistance be very effective.

3. He was also inquiring throughout Europe to interested scientists.

4. The remote-viewing experiments were conducted independently, and the results reported in their book *Mind Reach*.

5. The results of the test were later verified by a number of scientists in various laboratories around the country.

6. Professor John Hasted, chairman of the department of physics at Birbeck College in London, also conducted extensive experiments with such children in England, as did physicist Ted Bastin. Both found numerous children who could bend the metal without any physical contact.

Chapter 14

1. In an excellent book, *The Structure of Scientific Revolutions*, Thomas Kahn discusses precisely how paradigm shifts occur in the face of neglected or denied *experimental* results.

2. After the Geller work, I was asked to brief the then-director of the CIA, Ambassador (subsequently President) George H.W. Bush, on our activities and their results. In later years, I met with several Russian scientists who not only had documented results similar to ours, but also were actively using "psychic" techniques against the U.S. and its allies during the Brezhnev period.

3. In England, Professor John Hasted also witnessed children who were capable of this phenomenon, which he likened to "quantum tunnel effects." The children, he reported, appeared to teleport test objects into and out of sealed laboratory containers under controlled conditions on 12 occasions in one series of experiments.

4. Noted physicists from the United States, Britain, Denmark, France, and Germany all successfully tested Geller and/or "Geller children." They congregated in Iceland in 1977 to report their results and compare theories. Being unable to publish in professional journals, they produced a private volume of their findings, entitled *The Iceland Papers*, with a foreword by Nobel physicist, Brian Josephson.

Chapter 15

1. Today, nearly every freshman textbook in physics discusses the now-classic double-slit experiment that demonstrates wave/particle duality.

2. Physicists have for many years used the term *wavicles*. But this too is but an abstract label for energy, and labels are a product of mind.

Chapter 16

1. This is similar to a related paradoxical thought experiment called *Wigner's Friend*, which requires a succession of outside observers to collapse the wave function of each inner observer.

Chapter 18

1. Quasi-steady-state theorists, notably Fred Hoyle and Halton Arp, have amassed data and powerful new arguments against the big bang theory using modern telescopes. If correct, which it most likely is, arguments in this work are not affected, except the arrow of time must have a different origin.

2. The big bang theory, although a prevailing dogma in astronomy, is coming under increasing attack from those scientists who believe the concept of a quasi-steady state universe will eventually prevail.

3. Albert Einstein is often quoted on the relative (subjective) notion of time: "Sitting on a park bench with a pretty girl, the passing of one hour seems like one minute, while sitting on a hot stove for a few seconds may seem like several minutes have elapsed."

Chapter 19

1. A *dyad* is two different items inextricably linked, such as hot/cold, or the two faces of a coin.

2. Descartes' philosophy had a major influence on Western thinking because of his position within the Church. It also had some very negative consequences. He proposed that the body works as a machine, so it followed the laws of physics. The mind (or soul), on the other hand, is a nonmaterial entity and did not follow the laws of physics. Descartes also believed that only humans have minds, and that the mind interacts with the body through the pineal gland. Because he did not believe that animals had pineal glands, he proposed they could neither think nor feel. This led to the widespread use of the practice of vivisection throughout Europe.

3. The information we gather through our senses is quite incomplete. For instance, the unaided human eye only registers a tiny portion of the electromagnetic spectrum. Likewise, our other senses perceive only a fraction of the information available from their respective frequency domains. Collectively they sample but a tiny portion of the data and provide only a partial mapping, even with exquisite technologies.

Chapter 20

1. The superb studies of the late Joseph Campbell, reported in his classic *The Power of Myth*, eloquently speaks to many of these same cultural issues.

Chapter 21

1. I find Ken Wilber's magnificent commentary on the various types of inner experiences, *Spectrum of Consciousness* (1976), one of the best books of its kind.

2. I use "Self" in this work to imply the transcendent experience of Self as distinguished from the mundane experience of an ego self, separate and distinct from other objects of the world. "Self" should also not be construed as synonymous with traditional concepts of "soul," which implies an existence independent of the body, although it is tempting to make such an unwarranted extension.

3. I use the word *below* only as a convention to arbitrarily indicate direction, not in the sense of hierarchy. I've come to realize that the states of consciousness simply exist, none more lofty nor divine than others, but some definitely more coherent, acceptable, and satisfying in daily life.

Chapter 22

1. The terms *scholarly* and *mystical* used together imply the desire to be rational and coherently intuitive simultaneously.

2. Currently, the known Bose-Einstein condensates are low-temperature structures, and the understanding of them is still quite primitive.

Chapter 24

1. A theoretical model of how the zero-point field might accomplish this is described in Chapter 26, which describes the quantum hologram.

Chapter 25

1. If we could harness the energy of the zero-point field in space, unlimited energy could be available for travel that could approach the

speed of light. Travelers could have their local time relativistically extended so they would age several times more slowly than Earthlings. Practically speaking, however, the journeys would be one-way trips for the explorers, and final farewells for those left behind, if our current theories are correct.

2. This method of analysis was introduced in 1854, by mathematician Georg Bernhard Reimann. He set out to explore the abstract topology of curved spaces because Euclidean geometry couldn't properly account for such surfaces. His approach is called the *Riemann*, or metric tensor.

3. In his book *Hyperspace* (1994), Michio Kaku does a superb job of laying out the arguments and explanations for a 10-dimensional geometric interpretation of the universe, and a related cosmology that includes gravitational, electromagnetic, and quantum effects. The extra dimensions are on a scale smaller than nuclear diameters and require enormous energy to access.

Chapter 26

1. This phenomenon is called the *Berry phase*, or *geometric phase*, of a quantum system, and has been experimentally validated. This quantum mechanism associated with brain/body function was initially proposed by Peter Marcer in *World Futures* 44: 149–159.

2. Early work by Harold Puthoff (1980s), and more recent work by physicists Peter Marcer and Walter Schempp, reported in numerous papers (see bibliography), supports my belief that virtually all "psychic events" can be mathematically mapped as quantum exchanges in the brain between the zero-point field and macroworld objects.

3. Such radiation is of course hazardous, as it can pass through metal walls just as easily as it can one's body, damaging cell structure along the way.

4. Some of the most convincing evidence for an informational model of personality is the phenomenon of multiple personality disorder. The individual with MPD (often a result of an abusive childhood) has fragmented personalities that seem to completely manage the body independent of one another. They feel they are a different person, and

the external behaviors are quite different for each of the personalities; oftentimes physiological measures are different. Therapy consists of finding ways to integrate the various fragmented personalities into a more holistic pattern, accompanied by a comfortable internal rationale for doing so, which usually involves confronting and accepting the terrifying reality of the childhood abuse.

Chapter 28

1. Inertia must be resident in the zero-point field to prevent chaos in the movement of matter.

2. *Eidetic imagery* is unusually clear and vivid mental recall of material previously seen, the written in particular.

3. We often become more sensitive to our intuition when under emotional stress. Police often warn potential victims of crime to be aware of unusual inner sensations—they are often indicators that danger is lurking nearby.

Chapter 30

1. Recent discoveries with the Hubble telescope suggest new anomalies and controversy on the relative ages of the universe, certain galaxies, and even some stars. Evidence also has emerged that large red shifts occur in nearby galaxies, associated with creation of matter, creating doubt that the big bang is, in fact, a final theory, rather than a premature interpretation that red shift results only from recession.

2. I've had no personal encounters with UFOs, though I wish I had, so that I could speak from firsthand experience. I have, however, met with credible professionals within two governments who have testified to their own firsthand experiences with "close encounters" during their official duties. A wealth of classified information on the subject resides in military and intelligence files, which, in my opinion, should be released to the public.

Bibliography

Abbott, Edwin A. *Flat-Land*. New York: HarperCollins, 1983.

Arp, Halton. *Seeing Red: Redshifts, Cosmology and Academic Science*. Montreal: Aperion, 1998.

Barrow, John D. *The World Within the World*. New York: Oxford University Press, 1988.

Barrow, John D., and Frank J. Tipler. *The Anthropic Cosmological Principle*. Oxford: Oxford University Press, 1986.

Bentov, Itzhak. *Stalking the Wild Pendulum*. Rochester, Vt.: Destiny Books, 1977.

Bohm, David. *Wholeness and the Implicate Order*. New York: Ark Paperbacks, 1983.

Boslough, John. *Masters of Time: Cosmology at the End of Innocence*. Reading, Mass.: Addison Wesley, 1992.

Briggs, John P., PhD, and F. David Peat, PhD. *Looking Glass Universe: The Emerging Science of Wholeness*. New York: Simon & Schuster, 1984.

Butterworth, Eric. *The Universe Is Calling*. San Francisco: HarperCollins, 1993.

Casti, John L. *Complexification*. New York: HarperCollins, 1994.

———. *Searching for Certainty*. New York: William Morrow & Co., 1990.

Cohen-Tannoudji, Gilles. *Universal Constants in Physics*. New York: McGraw-Hill Inc., 1993.

Darling, David. *Equations of Eternity*. New York: Hyperion, 1993.

Davies, Paul. *About Time: Einstein's Unfinished Revolution*. New York: Simon & Schuster, 1995.

———. *God and the New Physics*. New York: Simon & Schuster, 1983.

———. *The Last Three Minutes*. New York: Basic Books, 1994.

———. *The Mind of God*. New York: Simon & Schuster, 1982.

Davies, Paul, and John Gribbon. *The Matter Myth*. New York: Simon & Schuster, 1992.

Davies, P.C.W., and Julian Brown. *Superstrings: A Theory of Everything?* Cambridge: Cambridge University Press, 1989.

Dyson, Freeman. *Infinite in All Directions*. New York: Harper & Row, 1988.

Edelglass, Stephen, George Maier, Hans Gebert, and John Davy. *Matter and Mind: Imaginative Participation in Science*. Herndon, Va.: Lindisfarne Books, 1992.

Gell-Mann, Murray. *The Quark and the Jaguar*. New York: W.H. Freeman & Co., 1994.

Gleick, James. *Chaos*. New York: Viking, 1987.

Goswami, Amit, PhD. *The Self-Aware Universe*. New York: G.P. Putnam's Sons, 1993.

Greenstein, George. *The Symbiotic Universe*. New York: William Morrow, 1988.

Gribbin, John. *In Search of the Big Bang*. Toronto: Bantam Books, 1986.

———. *Schrödinger's Kittens and the Search for Reality: Solving the Quantum Mysteries*. Boston: Little, Brown & Co., 1995.

Hawking, Stephen. *Black Holes and Baby Universes, and Other Essays*. Toronto: Bantam Books, 1993.

———. *A Brief History of Time*. Toronto: Bantam Books, 1988.

Herbert, Nick. *Elemental Mind*. New York: Dutton Books, 1993.

———. *Faster Than Light: Superluminal Loopholes in Physics*. New York: Penguin Group, 1988.

Hoyle, Fred, Geoffrey Burbidge, and Jayant Narlikar. *A Different Approach to Cosmology*. Cambridge: Cambridge University Press, 2000.

Kafotos, Menas, and Robert Nadeau. *The Conscious Universe*. Berlin, Germany: Spring-Verlag, 1990.

Kaku, Michio. *Hyper Space*. Oxford: Oxford University Press, 1994.

Lewin, Roger. *Complexity*. New York: MacMillan Publishing, 1992.

Mishlove, Jeffrey. *The Roots of Consciousness*. Tulsa, Okla.: Council Oak Books, 1993.

Morris, Richard. *Cosmic Questions, Galactic Halos, Cold Dark Matter, and the End of Time*. New York: John Wiley & Sons, 1993.

Overbye, Dennis. *Lonely Hearts of the Cosmos*. New York: Harper Collins, 1991.

Peat, F. David. *Einstein's Moon, Bell's Theorem, and the Curious Quest for Quantum Reality*. Chicago: Contemporary Books, 1990.

———. *The Philosopher's Stone*. Toronto: Bantam Books, 1991.

———. *Synchronicity: The Bridge Between Matter and Mind*. Toronto: Bantam Books, 1988.

Penrose, Roger. *The Emperor's New Mind*. New York: Penguin Books, 1989.

Prigogine, Ilya. *Order out of Chaos*. New York: Bantam Books, 1984.

Riordan, Michael, and David N. Schramm. *The Shadows of Creation, Dark Matter and the Structure of the Universe*. New York: W.H. Freeman & Co., 1991.

Rosen, Joe. *The Capricious Cosmos*. New York: MacMillan Publishing Co., 1991.

Rubik, Beverly, ed. "The Interrelationship Between Mind and Matter." Proceedings of the Center for Frontier Sciences, 1989.

Russell, Peter. *The White Hole in Time*. San Francisco: HarperCollins, 1992.

Sheldrake, Rupert. *Seven Experiments That Could Change the World*. London: Fourth Estate, 1994.

Smoot, George, and Keay Davidson. *Wrinkles in Time*. New York: Avon Books, 1993.

Stewart, Ian. *Does God Play Dice?* Cambridge, Mass.: Blackwell, 1992.

Talbot, Michael. *The Holographic Universe*. New York: Harper Collins, 1991.

———. *Mysticism and the New Physics*. New York: Penguin Books, 1981.

Thorne, Kip S. *Black Holes and Time Warps*. New York: W.W. Norton, 1994.

Tipler, Frank J. *The Physics of Immortality*. New York: Doubleday, 1994.

Van Flandern, Tom. "What the Global Positioning System Tells Us About Relativity." In *Open Questions in Relativistic Physics*, edited by Franco Selleri. Montreal: Apeiron, 1998.

Waldrop, M. Mitchell. *Complexity: The Emerging Science at the Edge of Order and Chaos*. New York: Simon & Schuster, 1992.

Weinberg, Steven. *The First Three Minutes*. New York: Basic Books, Inc., 1977.

Wolf, Fred Alan. *Parallel Universes*. New York: Simon & Schuster, 1988.

———. *Star Wave: Mind Consciousness and Quantum Physics*. New York: Macmillan Publishing, 1984.

Zajone, Arthur. *Catching the Light*. New York: Bantam Books, 1993.

Quantum Holography

Anandan, J. "The Geometric Phase." *Nature* 360, no. 26 (1992): 307–313.

Berry, M.V. "The Geometric Phase." *Scientific American* (December 1988): 26–32.

Bouwmeester, D., et al. *Nature* 390 (December 1997): 575–579.

Cramer, J.G. *Physics Review* 22, no. 362 (1980).

Eberhard, P.H. "Bell's Theorem without Hidden Variables." *Il Nuovo Cimento* 38, BI (March 1977): 75.

Frieder, B. Roy. *Physics from Fisher Information*. Cambridge: Cambridge University Press, 1998.

Haisch, B., A. Rueda, and H.E. Puthoff. "Physics of the Zero-Point Field. Implications for Inertia, Gravity, and Mass." *Speculations in Science and Technology* 20 (1997): 99–114.

———. "Advances in the Proposed Zero-Point Field Theory of Inertia." 34th AIAA Joint Propulsion Conference. *AIAA* 98 (1998): 31–43.

Hammeroff, S.R. "Quantum Coherence in Microtubules: A Neural Basis for Emergent Consciousness?" *Journal of Consciousness Studies* 1 (1994): 91–118.

Marcer, P.J. "Getting Quantum Theory off the Rocks: Nature as We Consciously Perceive It, Is Quantum Reality!" Proc. 14 International Congress of Cybernetics, Namur, Aug 21–25, 1995.

———. "The Jigsaw, the Elephant and the Lighthouse." Proceedings of ANPA 20 (1998).

Mercer, P.J., and W. Schempp. "The Brain as a Conscious System." *International Journal of General Systems* (1998).

———. "A Mathematically Specified Template for DNA and the Genetic Code in Terms of the Physically Realizable Processes of Quantum Holography." The Greenwich symposium on Living Computers, edited by A.M. Fedorec and P.J. Mercer (1996): 45–62.

———. "Model of the Neuron Working by Quantum Holography." *Informatica* 21 (1997): 519–534.

————. "The Model of the Prokaryote Cell as an Anticipatory System Working by Quantum Holography." *International Journal of Computing Anticipatory Systems. CHAOS* 2 (1998): 307–313.

Mitchell, Edgar. "Nature's Mind: the Quantum Hologram." *International Journal of Computing Anticipatory Systems* 7 (2000): 295.

Pribram, K.H. "Quantum Holography: Is It Relevant to Brain Function?" *Information Sciences* 115 (1999): 97–102.

Resta, R. "The Berry Phase." *Europhysics News* 28, no.19 (1997).

Schempp, W. "Harmonic Analysis on the Heisenberg Group with Application in Signal Theory." Pitman Research Notes in *Mathematics*, Series 14, Longman Scientific and Technica, London (1986).

————. "Quantum Holography and Neurocomputer Architectures." *Journal of Mathematical Imaging and Vision* 2 (1992): 109–164.

Sudbury, T. *Nature* 390 (December 1997): 551–552.

Death and Survival of Consciousness

Atwater, P.M.H. *Beyond the Light.* Secaucus, N.J.: Carol Publishing Group, 1994.

Darling, David. *Soul Search: A Scientist Explores the Afterlife.* New York: Villard Books, 1995.

Eadie, Betty J. *Embraced by the Light.* Placerville, Calif.: Gold Leaf Press, Placerville, 1992.

Grof, Stanislov, and Christina Grof. *Beyond Death.* New York: Thames & Hudson Inc., 1990.

Grosz, Anton. *Letters to a Dying Friend.* Wheaton, Ill.: Quest Books, 1989.

Harpur, Tom. *Life After Death.* Toronto: McClelland & Stewart Inc., 1993.

Holzer, Hans. *Life Beyond.* New York: McGraw-Hill, 1994.

Liverziani, Filippo. *Life, Death and Consciousness: Experiences Near and After Death.* New York: Avery Publishing Group, 1992.

Lorimer, David. *Whole in One: The Near-Death Experience and the Ethic of Interconnectedness.* London: The Penguin Group, 1990.

Lund, David H. *Death and Consciousness.* New York: Ballantine Books, 1989.

Meek, George W. *After We Die: What Then?* Columbus, Ohio: Ariel Press, 1987.

Moody, Raymond A., Jr., MD. *The Light Beyond*. New York: Bantam Books, 1988.

———. *Reunions*. New York: Villard Books, 1993.

Morse, Melvin, MD. *Closer to the Light: Learning from the Near-Death Experiences of Children*. East Sussex, UK: Ivy Books, 1990.

Morse, Melvin, MD, and Paul Perry. *Parting Visions: Uses and Meanings of Pre-Death, Psychic, and Spiritual Experiences*. New York: Villard Books, 1994.

———. *Transformed by the Light*. New York: Villard Books, 1992.

Perrish-Harra, Reverend Carol W. *The New Age Handbook on Death and Dying*. Santa Monica, Calif.: IBS Press, 1982.

Ring, Kenneth. *Heading Towards Omega*. New York: William Morrow & Co., 1984.

———. *The Omega Project*. New York: William Morrow & Co., 1992.

Rinpoche, Sogyal. *The Tibetan Book of Living and Dying*. San Francisco: HarperCollins, 1992.

Viney, Geoff. *Surviving Death*. New York: St. Martins Press, 1993.

Waiter, Carlos, MD. *Soul Remembers*. Sedona, Ariz.: Light Technology Publishing, 1992.

Whitton, Joel L., MD, PhD. *Joe Fisher: Life Between Life*. New York: Warner Books, 1986.

Woolger, Roger J., PhD. *Other Lives, Other Selves*. New York: Bantam Books, 1988.

Brain, Mind, Consciousness

Bateson, Gregory. *Mind and Nature: A Necessary Unity*. New York: Bantam Books, 1979.

Beakley, Brian, and Peter Ludlow, eds. *The Philosophy of Mind*. Cambridge, Mass.: MIT Press, 1992.

Blakeslee, Thomas R. *The Right Brain: A New Understanding of the Unconscious Mind and Its Creative Powers*. New York: Playboy Paperbacks, 1980.

Bucke, Richard Maurice, MD. *Cosmic Consciousness*. Secaucus, N.J.: Citadel Press, 1977.

Crick, Francis. *The Astonishing Hypotheses: The Scientific Search for the Soul*. New York: Charles Scribner's Son, 1994.

Cytowic, Richard E., MD. *The Man Who Tasted Shapes*. New York: G.P. Putnam's & Sons, 1993.

Damasio, Antonio R. *Descartes' Error: Emotion, Reason, and the Human Brain*. New York: G.P. Putnam's Sons, 1994.

Damiani, Anthony. *Looking into Mind*. Burden, N.Y.: Larson Publications, 1990.

Davidson, John. *The Formative Mind*. Rockport, Mass.: Element Inc, 1991.

Dennett, Daniel C. *Consciousness Explained*. Boston: Little, Brown & Co., 1991.

Denton, Dr. Derek. *The Pinnacle of Life: Consciousness and Self-Awareness in Humans and Animals*. San Francisco: HarperCollins, 1993.

Donaldson, Margaret. *Human Minds: An Exploration*. New York: Penguin Books, 1992.

Edelglass, Stephen, Georg Maier, Hans Gebert, and John Davy. *Matter and Mind: Imaginative Participation in Science*. Herndon, Vt.: Lindisfarne Books, 1992.

Edelman, Gerald M. *Bright Air, Brilliant Fire: On the Matter of the Mind*. New York: HarperCollins, 1992.

Falk, Dean. *Braindance*. New York: Henry Holt & Co., 1992.

Fezler, William, PhD. *Creative Imagery*. New York: Simon & Schuster, 1989.

Gardner, Howard. *Frames of Mind: The Theory of Multiple Intelligences*. New York: HarperCollins, 1993.

Gazzaniga, Michael S. *Mind Matters*. Boston: Houghton Mifflin, 1988.

Goleman, Daniel, and Richard J. Davidson. *Consciousness: Brain States of Awareness and Mysticism*. New York: Harper & Row, 1979.

Griffin, Donald R. *Animal Minds*. Chicago: University of Chicago Press, 1992.

Grof, Stanislav. *Beyond the Brain*. Albany, N.Y.: State University of New York, 1995.

Grof, Stanislav, and Hal Zina Bennett, PhD. *The Holotropic Mind: The Three Levels of Human Consciousness and How They Shape Our Lives*. San Francisco: HarperCollins, 1992.

Harth, Erich. *The Creative Loop: How The Brain Makes a Mind*. Boston: Addison-Wesley Publishing Co., 1993.

Harman, Willis, PhD, and Howard Reingold. *Higher Creativity: Liberating the Unconscious for Breakthrough Insights*. New York: G.P. Putnam's Sons, 1984.

Harman, Ernest. *Boundaries of the Mind: A New Psychology of Personality*. New York: Basic Books, 1991.

Hauper, Judith, and Dick Teresi. *The Three Pound Universe*. Los Angeles: Jeremy P. Tarcher, Inc., 1986.

Howe, M.J.A. *Fragments of Genius*. London: Routledge, 1989.

Jahn, Robert C., and Brenda J. Dunne. *Margins of Reality: The Role of Consciousness in the Physical World*. San Diego: Harcourt Brace Jovanovich, 1987.

Johnson, George. *In the Palaces of Memory*. New York: Vintage Books, 1992.

Kosslyn, Stephen M., and Oliver Koenig. *Wet Mind: The New Cognitive Neuroscience*. New York: The Free Press/Macmillan Inc., 1992.

La Berge, Stephen, PhD, and Howard Rheingold. *Exploring the World of Lucid Dreaming*. New York: Ballantine Books, 1990.

Laughlin, Charles D., Jr., John McManus, and Eugene G. d'Aquily. *Brain, Symbol and Experience*. Boston: Shambhala, 1990.

Lockwood, Michael, and Basil Blackwood. *Mind, Brain, and the Quantum*. Oxford: Oxford University Press, 1989.

Nadeau, Robert L. *Mind, Machines, and Human Consciousness: Are There Limits to Artificial Intelligence?* Chicago: Contemporary Books, 1994.

Ornstein, Robert. *The Evolution of Consciousness*. New York: Prentice Hall, 1991.

Penrose, Roger. *Shadows of the Mind*. Oxford: Oxford University Press, 1994.

Pinker, Steven. *The Language of Instinct*. New York: William Morrow & Co., 1994.

Restak, Richard, MD. *The Brain Has a Mind of Its Own*. New York: Crown Publishers, 1991.

———. *The Brain: The Last Frontier*. Garden City, N.Y.: Doubleday & Co., 1979.

Rose, Steve. *The Conscious Brain*. New York: Paragon House, 1973.

Rosenfield, Israel. *The Invention of Memory*. New York: Basic Books, Inc., 1988.

Searle, John R. *The Rediscovery of the Mind*. Cambridge, Mass.: MIT Press, 1992.

Silver, Todd. *Breaking the Mind Barrier*. New York: Simon & Schuster, 1990.

Smith, Adam. *Powers of Mind*. Toronto: Ballantine Books, 1975.

Stine, Jean, and Camden Benares. *It's All in Your Head*. New York: Prentice Hall, 1994.

Tart, Charles T., PhD. *Open Mind, Discriminating Mind*. San Francisco: Harper & Row, 1989.

Wills, Christopher. *The Runaway Brain*. New York: Basic Books, 1993.

Wolf, Fred Alan, PhD. *The Dreaming Universe*. New York: Simon & Schuster, 1994.

Philosophy and Metaphysics

Abraham, Ralph, Terence McKenna, and Rupert Sheldrake. *Trialogues at the Edge of the West*. Santa Fe, N.M.: Bear & Co., 1992.

Appleyard, Bryan. *Understanding the Present Science and the Soul of Modern Man*. New York: Doubleday, 1992.

Bache, Christopher M. *Dark Night, Early Dawn*. Albany, N.Y.: State University of New York Press, 2000.

Barrow, John D. *Theories of Everything*. New York: Fawcett, 1991.

Batson, Gregory. *Steps to an Ecology of Mind*. New York: Ballantine Books, 1972.

Capra, Fritjof. *The Tao of Physics*. Boston: Shambhala, 1991.

———. *The Turning Point*. Toronto: Bantam Books, 1982.

Capra, Fritjof, and David Steindl-Rast. *Belonging to the Universe*. San Francisco: HarperCollins, 1991.

Casti, John L. *Paradigms Lost*. New York: Avon Books, 1989.

Chopra, Deepak, MD. *Unconditional Life*. New York: Bantam Books, 1991.

Combs, Allan, and Mack Holland. *Synchronicity, Science, Myth, and the Trickster*. New York: Paragon House, 1990.

Ferris, Timothy. *The Mind's Sky, Human Intelligence in a Cosmic Context*. New York: Bantam Books, 1992.

Gribbin, John, and Martin Rees. *Cosmic Coincidences: Dark Matter, Mankind, and Anthropic Cosmology*. New York: Bantam Books, 1989.

Harman, Willis, PhD. *Global Mind Change*. Indianapolis, Ind.: Knowledge Systems, Inc., 1988.

Harman, Willis, and Jane Clark. *New Metaphysical Foundation of Modern Science*. Sausalito, Calif.: Institute of Noetic Sciences, 1994.

Jastrow, Robert. *God and the Astronomers*. New York: W.W. Norton & Co., 1992.

Krishnamurti, J., and Dr. David Bohm. *The Ending of Time*. San Francisco: HarperCollins, 1985.

Kuhn, Thomas S. *The Structure of Scientific Revolutions*. Chicago: University of Chicago Press, 1970.

Martin, Graham Dunston. *Shadows in the Cave*. New York: Penguin Books, 1990.

Morris, Richard. *The Edges of Science*. New York: Prentice Hall Press, 1990.

Nagel, Thomas. *Mortal Questions*. Cambridge, Mass.: Cambridge University Press, 1979.

Ornstein, Robert, and Paul Ehrlich. *New World, New Mind*. New York: Simon & Schuster Inc. 1989.

Pearce, Joseph Chilton. *The Crack in the Cosmic Egg*. New York: Julian Press, 1988.

Rucker, Rudy. *The Fourth Dimension*. Boston: Houghton Mifflin, 1984.

Sheldrake, Rupert. *The Presence of the Past*. New York: Vintage Books, 1988.

———. *The Rebirth of Nature*. New York: Bantam Books, 1991.

Talbot, Michael. *Beyond the Quantum*. New York: Bantam Books, 1987.

———. *The Holographic Universe*. New York: HarperCollins, 1991.

Tart, Charles T. *Waking Up: Overcoming the Obstacles to Human Potential*. Boston: Shambhala Publications, 1986.

Thagard, Paul. *Conceptual Revolutions*. Princeton, N.J.: Princeton University Press, 1992.

Theobald, Robert. *The Rapids of Change*. Indianapolis, Ind.: Knowledge Systems, 1987.

Weber, Renee. *Dialogues with Scientists and Sages*. London: Penguin Group, 1990.

Wilber, Ken. *Eye to Eye: The Quest for a New Paradigm*. Boston: Shambhala, 1990.

———. *The Holographic Paradigm and Other Paradoxes*. Boston: Shambhala, 1982.

———. *The Spectrum of Consciousness*. Quest Books: Wheaton, Ill., 1993.

Zohar, Danalt, and Ian Marshall. *The Quantum Society: Mind, Physics and a New Social Vision*. New York: William Morrow & Company, Inc., 1994.

Mind/Body Connection

Becker, Robert O., MD, and Gary Selden. *The Body Electric*. New York: William Morrow Publishers, 1985.

Borysenko, Joan, PhD. *Minding the Body, Mending the Mind*. Reading, Mass.: Addison-Wesley, 1987.

Byrd, R.C. "Positive Therapeutic Effect of Intercessory Prayer in a Coronary Care Population." *Southern Medical Journal* 81(7): 826–829 (1988).

Chopra, Deepak, MD. *Ageless Body, Timeless Mind*. New York: Harmony Books, 1993.

———. *Quantum Healing*. New York: Bantam Books, 1990.

Chopra, Deepak, MD, and Norman Cousins. *Anatomy of an Illness*. New York: Bantam Books, 1985.

Dacher, Elliott S., MD. *PNI, the New Mind/Body Healing Program*. New York: Paragon House, 1993.

Dienstfrey, Harris. *Where the Mind Meets the Body*. New York: Harper Perennial, 1991.

Dossy, Larry, MD. *Healing Words*. San Francisco: HarperCollins, 1993.

———. *Meaning and Medicine*. New York: Bantam Books, 1991.

———. *Space, Time, and Medicine*. Boston: Shambhala, 1985.

Epstein, Gerald, MD. *Healing Visualizations*. New York: Bantam Books, 1989.

Franklin, Jon. *Molecules of the Mind*. New York: Dell Publishing Co., 1987.

Garfield, Patricia, PhD. *Healing Power of Dreams*. New York: Simon & Schuster, 1991.

Goleman, Daniel, PhD, and Joel Gurin, eds. *Mind Body Medicine*. Yonkers, N.Y.: Consumer Reports Books, 1993.

Green, Elmer, and Alyce Green. *Beyond Biofeedback*. Ft. Wayne, Ind.: Knoll Publishing Co., 1989.

Hodgkinson, Neville. *Will to Be Well*. York Beach, Maine: Samuel Weiser, Inc., 1984.

Institute of Noetic Sciences. *The Heart of Healing*. Atlanta, Ga.: Turner Publishing, 1993.

Justice, Blair, PhD. *Who Gets Sick*. New York: Jeremy P. Tarcher, 1987.

Moskowitz, Reed C. *Your Healing Mind*. New York: Avon Books, 1992.

Ornstein, Robert, PhD, and David Sobel, M.D. *Healing Brain*. New York: Simon & Schuster, 1987.

———. *Healthy Pleasures*. Reading, Mass.: Addison Wesley, 1989.

Rossman, Martin L. *Healing Yourself*. New York: Pocket Books, 1989.

Siegel, Bernie S., MD. *Peace, Love, and Healing*. New York: Harper & Row, 1989.

Spirituality and Philosophy

Borysenko, Joan, PhD. *Fire in the Soul*. New York: Warner Books, 1993.

Capra, Fritjof. *Uncommon Wisdom*. New York: Bantam Books, 1988.

Harner, Michael. *The Way of the Shaman*. San Francisco: HarperCollins, 1990.

Hastings, Arthur. *With the Tongues of Men and Angels*. Fort Worth, Tx.: Holt, Rinehart & Winston, 1991.

Hofstadter, Douglass R., and Daniel C. Dennett. *The Mind's I*. Toronto: Bantam Books, 1982.

Huxley, Aldous. *The Perennial Philosophy*. New York: Harper & Row, 1970.

Kalweit, Holger. *Shamans, Healers, and Medicine Men*. Boston: Shambhala, 1992.

Keys, Ken, Jr. *Handbook to Higher Consciousness*. Coos Bay, Ore.: Living Love Publications, 1975.

Monroe, Robert A. *Ultimate Journey*. New York: Doubleday, 1994.

Oyle, Irving, and Susan Jean. *The Wisdom Within*. Tiburon, Calif.: H.J. Kramer, Inc., 1992.

Ravindra, Ravi. *Science and Spirit*. New York: Paragon House, 1991.

Spalding, Baird T. *Life and Teaching of the Masters of the Far East, Vols. 1–5*. Camarillo, Calif.: DeVorss & Co., 1924.

Tart, Charles T. *Living the Mindful Life*. Boston: Shambhala, 1994.

———. *Transpersonal Psychologies: Perspectives on the Mind from Seven Great Spiritual Traditions*. San Francisco: HarperCollins, 1992.

Vaughan, Frances E. *Awakening Intuition*. New York: Anchor Books Doubleday, 1979.

Watts, Alan. *The Way of Zen*. New York: Vintage Books, 1957.

Evolution, Cognitive Science

Abraham, Ralph. *Chaos Gaia Eros*. San Francisco: HarperCollins, 1994.

Asimov, Isaac. *Beginnings*. New York: Berkley Books, 1989.

Bloom, Howard. *The Lucifer Principle*. New York: The Atlantic Monthly Press, 1995.

Calvin, William H. *The Ascent of Mind*. New York: Bantam Books, 1991.

Carey, Ken. *The Third Millennium*. San Francisco: HarperCollins, 1991.

Cornwell, John, ed. *Nature's Imagination*. Oxford: Oxford University Press, 1995.

Dawkins, Richard. *The Blind Watchmaker*. New York: W.W. Norton & Company, 1987.

Donald, Merlin. *Origins of the Modern Mind*. Cambridge, Mass.: Harvard University Press, 1991.

Dozier, Rush W., Jr. *Codes of Evolution*. New York: Crown Publishers, Inc., 1992.

Duve, Christian de. *Vital Dust: Life as a Cosmic Imperative*. New York: Basic Books/Harper Collins, 1995.

Edey, Maitland A., and Donald C. Johnson. *Blueprints: Solving the Mystery of Evolution*. Boston: Little, Brown & Co., 1989.

Gardner, Howard. *The Mind's New Science*. New York: HarperCollins, 1987.

Gifford, Don. *The Farther Shore*. New York: First Vintage Books, 1991.

Gribbon, John. *In the Beginning*. New York: Back Bay Books. 1993.

Harris, Marvin. *Our Kind*. New York: Harper & Row, 1989.

Henderson, Hazel. *Paradigms in Progress*. Indianapolis, Ind.: Knowledge Systems, Inc., 1991.

Johnston, Charles M., MD. *Necessary Wisdom*. Seattle: ICD Press, 1991.

Korten, David C. *Getting to the 21st Century*. Bloomfield, Conn.: Kumarian Press, 1990.

Layzer, David. *Cosmogenesis*. New York: Oxford University Press, 1990.

Litrak, Stuart, and A. Wayne Senzee. *Towards a New Brain*. Englewood Cliffs, N.J.: Prentice Hall, 1986.

Maturana, Humberto R., PhD, and Francisco J. Varela, PhD. *The Tree of Knowledge: The Biological Roots of Human Understanding.* Boston: Shambhala, 1992.

Moravec, Hans. *Mind Children: The Future of Robot and Human Intelligence.* Cambridge, Mass.: Harvard University Press, 1988.

Murphy, Michael. *The Future of the Body.* Los Angeles: Jeremy P. Tarcher, Inc., 1992.

Paepke, C. Owen. *The Evolution of Progress.* New York: Random House, 1993.

Pearse, Joseph Chilton. *Evolution's End.* San Francisco: HarperCollins, 1992.

Rao, K. Ramakrishna, ed. *Cultivating Consciousness.* Westport, Conn.: Praeger, 1993.

Sahtouris, Elisabet. *Gaia: The Human Journey from Chaos to Cosmos.* New York: Pocket Books, 1989.

Swimme, Brian, and Thomas Berry. *The Universe Story.* San Francisco: HarperCollins, 1992.

Taylor, Gordon Ratray. *The Natural History of the Mind.* New York: E.P. Dutton, 1979.

Thomas, Lewis. *The Fragile Species.* New York: Macmillan Publishing, 1992.

Van Doren, Charles. *A History of Knowledge.* New York: Ballantine Books, 1991.

Varela, Francisco J., Evan Thompson, and Eleanor Rosch. *The Embodied Mind.* Cambridge, Mass.: MIT Press, 1991.

Wheatley, Margaret J. *Leadership and the New Science.* San Francisco: Berrett-Koehler Publishers, 1992.

Wills, Christopher. *The Wisdom of the Genes.* New York: Basic Books, 1989.

Young, Louise B. *The Unfinished Universe.* Oxford: Oxford University Press, 1986.

UFOs, Parapsychology, and Psychic Phenomena

Bova, Ben, and Byron Preiss, eds. *First Contact.* London: Penguin, 1991.

Drake, Frank, and Dava Sobel. *Is Anyone Out There?* New York: Delacorte Press, 1992.

Dossey, L. *Healing Words.* San Francisco: HarperCollins, 1993.

Dunne, B.J., R.D. Nelson, and R.G. Jahn. "Operator-Related Anomalies in a Random Mechanical Cascade." *JSE* 2:155-80 (1988).

Mitchell, Janet Lee, PhD. *Out-of-Body Experiences*. New York: Ballantine Books, 1981.

Monroe, Robert A. *Far Journeys*. New York: Doubleday, 1985.

Puharich, Andrija, MD, LLD, ed. *Iceland Papers*. Amherst, Wisc.: Essentia Research Association, 1979.

Puthoff, H.E. "CIA-Initiated Remote Viewing Program at Stanford Research Institute." *JSE* 10 (1996): 63–76.

Puthoff, H.E., and R. Targ. "A Perceptual Channel for Information Transfer over Kilometer Distances: Historical Perspective and Recent Research." Proceedings of the IEEE 64 (1976): 329–354.

Radin, Dean. *The Conscious Universe*. San Francisco: HarperCollins, 1997.

Randle, Kevin D. *UFO Casebook*. Clayton, Australia: Warner Books, 1989.

Randle, Kevin D., and Donald R. Schmitt. *UFO Crash at Roswell*. New York: Avon Books, 1991.

Rao, K. Ramakrishna, ed. *The Basic Experiments in Parapsychology*. Jefferson, N.C.: McFarland & Co., 1984.

Schlitz, M., and R. Wiserman. "Experimenter Effects and the Remote Detection of Staring." *Journal of Parapsychology* 61 (September, 1997).

Schmeidler, G.R., and J.G. Craig. "Moods and ESP Scores in Group Testing." *Journal of ASPR* 66 (3): 280–287 (1972).

Tiller, A.W. *Science and Human Transformation*. Walnut Creek, Calif.: Pavior Publishing, 1997.

Utts, J.M. "Replication and Meta-Analysis in Parapsychology." *Statistical Science* 6: 363–382 (1991).

Vallee, Jacques. *Confrontations*. New York: Ballantine Books, 1990.

———. *Revelations*. New York: Ballantine Books, 1991.

Weiss, Brian L., MD. *Many Lives, Many Masters*. New York: Simon & Schuster, Inc., 1988.

Ecology

Gore, Al. *Earth in the Balance: Ecology and the Human Spirit*. Boston: Houghton Mifflin, 1992.

Lerner, Steve. *Earth Summit*. Bolinas, Calif.: Common Knowledge Press, 1991.

Lovelock, J.E. *Gaia*. Oxford: Oxford University Press, 1991.

Meadows, Donella H., Dennis L. Meadows, and Jorgen Randers. *Beyond the Limits*. Post Mills, Vt.: Chelsea Green Publishing, 1992.

Space

Chaikin, Andrew. *A Man on the Moon: The Voyages of the Apollo Astronauts*. New York: Viking Press, 1994.

Mitchell, Edgar. "Space Flight as an Anticipatory Computing System." Conference Proceedings of Computing Anticipatory Systems (CASYS '99), American Institute of Physics.

Shepard, Alan, Deke Slayton, Jay Barbree, and Howard Benedict. *Moon Shot*. Nashville, Tenn.: Turner Publishing, 1995.

Index

C

Cardeña, Etzel, 267–268
Carnegie Mellon, 16
Carr, Jerry, 32
causality, 140
Cernan, Gene, 51, 57
Chaffee, Roger, 32
Challenger explosion, 32
Chalmers, David, 270–271
chaos theory, 150–152, 181
    atomic ordering and, 188
Chen, Norbu, 97–99, 109–110,
    112–113, 129, 209
Clarke, Arthur C., 37
collective unconscious concept, 210
Collins, Michael, 41
complementarity principle, 123–124
complexity theory, 150
    atomic ordering and, 188
computer technology, 40–45
Cone Crater, 58, 60, 63, 71
consciousness, 77–78, 83, 85–90, 183
    dualism and, 184–186
    quantum mechanics and, 119–120
    states of, 169–174
    wave equation effects of, 138–142
Cooper, Gordon, 42, 264
Copenhagen interpretation of
    quantum physics, 125, 132
Cosmic Consciousness (Bucke), 80, 81
creation process, 147
creativity, 240
cultural beliefs, 107
Curie, Marie, x
cybernetics, 143

D

Darwin, Charles, x, 107
da Vinci, Leonardo, x
death, 218–222

de Broglie, Louis, 122–123
decay process, 147
delayed choice experiment, 135
Descartes, René, x, 7–8, 107, 157, 175
d'Espagnat, Bernard, 270
determinism, volition and, 233
deterministic self-organizing
    (autopoietic) systems, 188
dimensionality, 201–204
distraction index, 191
Dornberger, Walter, 36
double-slit experiment, 135
Douglas, Kirk, 50
Drake, Frank, 38
dual-aspect monism, 270
dualism, 83, 107, 232–233
    consciousness and, 184–186
    healing, 194
    Self and, 172–174
Duke, Charlie, 79
dyadic model, 155–161, 270
    described, 156
    extraordinary human
        functioning, 192–196
    first-person insights, 169–179
    intelligent universe
        organisms, 187–191
    matter and, 158–159
    mind and, 159–161
    religion and, 162–168
    Self and, 169–179

E

Earth, 8, 55–56
ecstasy of unity, 4
Edison, Thomas, 186
Edwards Air Force Base Aerospace
    Research Pilots School, 26–27
$E=MC^2$ concept, 123
ego state, 173–174, 218

I

ineffable state, 170–171
information, 114, 115, 143–152
    energy patterns in nature and, 144
    far-from-equilibrium processes
        and, 145–147
    managing, intentionally, 143–144
    meaning and, 144
    nonlinear processes and, 150–152
    Shannon theory of, 143
    signal and, 144
    time and, 147–150
    Weiner scientific definition of, 143
Institute of Noetic Sciences
    (IONS), ix, 231, 263, 267
    board of directors, 89–90
    founding of, 5–6, 86
    funding for, 87–89
    purpose of, 86, 88
intelligent universe organisms, 187–191
intentionality, 91–95, 115,
    183–184, 243
intuition, emotion, and intellect
    understanding, 5
intuition sources, 239–240
intuitive information sources, 239–240
intuitive insights, 79
irreversible dissipative processes, 146
    time and, 147–150
Irwin, Jim, 79

J

Jahn, Robert, 243
James, William, 270
Jesus, 235
Johnson, Lyndon, 25
Johnson, Olaf, 72
Journal of Parapsychology, The, 72
Jung, Carl, 210

K

Kaluza, Theodor, 204
karma, 223
Kissinger, Henry, 50
Kittyhawk (command
    module), 3, 56, 236
knowing, 120–121, 127, 136, 209, 227
    consciousness and, 138–142

L

LaBerge, Stephen, 214–215
landing radar rumor, 62
language development, 94–95
Lao Tse, 94, 164, 249
learning, 180–182
    universe, 187–188
Leibniz, Gottfried, x
Lenov, Alexii, 79
Lorenz, Edward, 151
love, 225–226
Lovell, Jim, 41, 42, 43
Lovelock, James, 187
lucid dreaming, 214–215
lunar day, 64

M

Makoto Shichida, 238
Manned Orbiting Laboratory
    (MOL), 26
Many Worlds interpretation, 140–141
Marcel, Jesse, Jr., 264
Massachusetts Institute of
    Technology, 20–21, 22
Masters, Robert, 5
matrix algebra, 204–205
matter
    mind and, 158–161
    as particle and wave, 123
Mattingly, Ken, 32, 45
Maxey, Edward, 49, 58, 71, 72, 112

O

Oberth, Hermann, 35
objective *vs.* subjective events, 108–109
one-God concept, 165
order, nonlocality and, 187–188
Order Out of Chaos (Prigogine), 145
O'Regan, Brendan, 89
Osis, Karlis, 72
out-of-body experience (OBE), 213–214

P

Parr, Jack, 112
particles, 123, 125, 184
    attributes of, 127–129
Pauli, Wolfgang, 270
peak oil, 259
Penrose, Roger, 181–182
Perennial Wisdom, 182–184
persona, 193
Physics of Immortality, The
    (Tipler), 219, 253
Plan B 3.0 (Brown), 260
Planck, Max, 116, 122
Plato, 157
Platonic Idealism, 187
Platonic Idealist, 184
power of belief experiences, 96–105
precognition, 243
Prigogine, Ilya, 145–146, 150
process of creation, 147
process of decay, 147
psi missing result, 72
psychoactivity, 115
psychokinesis, 78, 100–101,
    103, 110, 119
psychological time, 149
psycho-navigation techniques, 162–163
Puharich, Andrija, 100–101
Puthoff, Harold, 101–102,
    111, 207, 237

Q

quantum eraser experiment, 135
quantum hologram (QH),
    131, 210–211, 213
quantum jumping, 112
quantum mechanical interpretations/
    paradoxes, 130–137
quantum mechanics, 119–129
    Cartesian duality and, 125–126
    consciousness and, 119–120
    knowing and, 120–121, 127
    light concepts and, 122–123
    matter as particle and
        wave and, 123
    principles/paradoxes causing
        confusion with, 123–124
    reality models and, 121–122
quantum physics, 107, 254
    Copenhagen interpretation
        of, 125, 132

R

Radin, Dean, 243
reality models, 121–122
regression therapy, 217
reincarnation of souls, 180, 215, 216
religion
    diadic model and, 162–168
    and mystical experience, 162
    psycho-navigation
        techniques, 162–163
    vs. science, 7–9, 48, 70–71
religious experiences, 80
Remington, Jim, 22–23, 47
remote viewing, 102, 213–214
resonance, 213
Rettig, Anita, 87, 112–113
Rhine, Joseph Banks, 48,
    57–58, 71, 107, 242
Rolfs, Henry, 101

About the Author

On January 31, 1971, Dr. Edgar Mitchell, then a U.S. Navy captain, embarked on a journey through outer space of some 500,000 miles that resulted in his becoming the sixth man to walk on the moon. That historic journey terminated safely nine days later on February 9, 1971, and was made in the company of two other men of valor: Admiral Alan Shepard and Colonel Stuart Roosa.

Scientist, test pilot, naval officer, astronaut, entrepreneur, author, and lecturer, Dr. Mitchell's extraordinary and varied career personifies humankind's eternal thrust to widen its horizons as well as explore its inner soul.

His academic background includes a bachelor of science in industrial management from Carnegie Mellon University in 1952, a bachelor of science in aeronautics from the U.S. Naval Postgraduate School in 1961, and a doctor of science degree in aeronautics and astronautics from the Massachusetts Institute of Technology in 1964. In addition, he has received four honorary doctorates, one from each of the following: New Mexico State University, the University of Akron, Carnegie Mellon University, and Embry-Riddle University.

In 1973, a year after retiring from the U.S. Navy and the astronaut program, Dr. Mitchell founded the Institute of Noetic Sciences. It is a foundation organized to sponsor research in the nature of consciousness. He is cofounder of the Association of Space Explorers, an international organization founded in 1984 for all who share the experience of space travel. Both organizations are educational organizations developed to provide new understanding of the human condition resulting from the epoch of space exploration.

He is the author of *Psychic Exploration: A Challenge for Science*, published by G.P. Putnam's Sons in 1974, a major reference book. He was a frequent guest on radio and television talk shows, and has been featured in several documentary films relative to his interests. Having retired from government service in 1972, Dr. Mitchell continued to write, speak, and conduct research for a number of new books. He was also a consultant to a limited number of corporations and foundations.

Dr. Mitchell's honors and awards include the Presidential Medal of Freedom, the USN Distinguished Service Medal, the NASA Distinguished Service Medal, the NASA Group Achievement Award (three times), and was a 2005 nominee for the Nobel Peace Prize.

Dr. Mitchell passed away February 4, 2016, in West Palm Beach, Florida, on the eve of the forty-fifth anniversary of his lunar landing.

From Dr. Mitchell's Writing

Suddenly, from behind the rim of the moon, in long, slow-motion moments of immense majesty, there emerges a sparkling blue and white jewel, a light, delicate, sky-blue sphere laced with slowly swirling veils of white, rising gradually like a small pearl in a thick sea of black mystery—it takes more than a moment to fully realize this is Earth—home.

On the return trip home, gazing through 240,000 miles of space toward the stars and the planet from which I had come, I suddenly experienced the universe as intelligent, loving, harmonious.

My view of our planet was a glimpse of divinity.

We went to the moon as technicians; we returned as humanitarians.